"RETAINED BY THE PEOPLE"

BICENTENNIAL ESSAYS ON THE BILL OF RIGHTS

Co-sponsored by Oxford University Press
and the Organization of American Historians

Kermit L. Hall, General Editor

EDITORIAL BOARD

Michal Belknap, Harold M. Hyman,
R. Kent Newmeyer, William M. Wiecek

FAIR TRIAL
*Rights of the Accused in
American History*
David J. Bodenhamer

THE GUARDIAN OF EVERY OTHER RIGHT
*A Constitutional History of
Property Rights*
James W. Ely, Jr.

PROMISES TO KEEP
*African Americans and the Constitutional Order,
1776 to the Present*
Donald G. Nieman

"RETAINED BY THE PEOPLE"
*A History of American Indians and
the Bill of Rights*
John R. Wunder

OTHER VOLUMES ARE IN PREPARATION

"RETAINED BY THE PEOPLE"

A History of American Indians and the Bill of Rights

JOHN R. WUNDER

New York   Oxford
OXFORD UNIVERSITY PRESS
1994

Oxford University Press

Oxford   New York   Toronto
Delhi   Bombay   Calcutta   Madras   Karachi
Kuala Lumpur   Singapore   Hong Kong   Tokyo
Nairobi   Dar es Salaam   Cape Town
Melbourne   Auckland   Madrid

and associated companies in
Berlin   Ibadan

Copyright © 1994 by Oxford University Press, Inc.

Published by Oxford University Press, Inc.,
200 Madison Avenue, New York, New York 10016

Oxford is a registered trademark of Oxford University Press

Library of Congress Cataloging-in-Publication Data
Wunder, John R.
''Retained by the people'' :
a history of American Indians and the Bill of Rights /
John R. Wunder.
p.   cm. — (Bicentennial essays on the Bill of Rights)
Includes bibliographical references and index.
ISBN 0–19–505562–4 (cloth).
ISBN 0–19–505563–2 (pbk.)
1. Indians of North America—Civil rights—History.
I. Title.
II. Series: Organization of American Historians bicentennial
essays on the Bill of Rights.
KF8205.W86   1994   342.73'0872—dc20   [347.302872]   93–9804

2 4 6 8 9 7 5 3 1

Printed in the United States of America
on acid-free paper

For
Susan
and the Legal Warriors

Preface

In today's world, the words of the past are often lost or misplaced. For many Native Americans, however, that is not usually the situation, as many know about themselves, their relatives, and their tribe. They also know that when their forebears, their great-grandfathers and great-grandmothers, lived on their traditional homelands, they called themselves The People.

The largest Indian nation today is the Navajo Nation, located in the Four Corners region of Arizona, New Mexico, and Utah. Navajo is not their name for themselves; rather, they call themselves Diné. And they live not in Navajoland, but in Dinetah. In Navajo, *Diné* means "The People." Like the Navajo, the vast majority of Indians today must respond to misnomers. They know that they are members of tribes whose native language refers to themselves as The People.

The Ninth Amendment to the United States Constitution is brief. Its words are powerful, but more often than not, judges, lawyers, legislators, and even presidents have ignored or disregarded them. Still, the words are worth repeating: "The enumeration in the Constitution, of certain rights, shall not be construed to deny or disparage others *retained by the people.*" Whether The People have retained certain rights is the subject of this book.

Three important clarifications must be made before progressing any further. In this book, several terms are used interchangeably, including "Indian," "Native American," "native peoples," and "indigenous peoples." This is intentional, as all these names have been adopted by one or more native organizations as a form of reference. To recognize one over any of the others therefore would not reflect current and past scholarship. The term "Indian" is another misnomer, but it is used so pervasively that to avoid it would puzzle the reader, which is why "Indian" is found in the body and subtitle of this book. My own preference is the term "native peoples," but it is used primarily in Canada and not in the United States, and rather than confuse potential readers, including library bibliographers, I have reluctantly opted for a term historically conceived out of ignorance. Similarly "tribe" and "nation" are interchanged.

The second point to be made regards specificity. This is a book about more than 300 different nations of North America. As the reader will quickly observe, it is extremely difficult to generalize about all Indians; nevertheless, the nature of this kind of project requires generalization. I have tried to compensate for the stereotyping or overgeneralizing by being as specific as possible in the examples. Moreover, much of Native American history is nameless. Here, too, I have erred on the side of specificity, noting individual contributions and specific tribal actions whenever possible.

A third issue involves the use of Indian fiction. In each chapter, references or allusions are made to Native American literature, for several reasons. Like Russian fiction, Indian fiction has the wonder of often being more meaningful and truthful than are many of the histories written about Indians. Much Native American prose is also forceful, succinct, and purposeful. Moreover, many modern Native American poets and novelists use themes about law, rights, and the past—all of which concern this book.

A word of caution is needed before turning the pages. By necessity, this extremely complex topic has been condensed. The original manuscript was pared in order to fit within the Bicentennial Essays series. This process required deleting extended annotated notes and legislative titles of statutes in the notes and deleting or shortening sections on Indian court jurisdiction, international law, and federal appellate court decisions. These topics and many others remain fruitful fields for investigation.

■   ■   ■

This book would not have been possible without the strong encouragement and support of numerous people. I am very grateful to Kermit Hall, the general editor; the members of his editorial advisory committee; and the Organization of American Historians for allowing and encouraging this topic to be explored in the Bicentennial Essays on the Bill of Rights series. Kermit's interest has been indispensable to this project's completion. The contributions of Willard Rollings, Rita Napier, Orlan Svingen, and James Ronda, who helped me think through the conceptualization of placing Indians and law in history, were and continue to be invaluable. A great deal is owed to my good friend Richard White, who always listens to my scholarly rantings and then tells me the truth.

The Indian communities of Lincoln and Nebraska were most helpful. Especially appreciated are the interest and conversations with history graduate students Susan Miller and Denny Smith; faculty from the Institute for Ethnic Studies, Charles Ballard and Cynthia Willis, at the University of Nebraska–Lincoln; Clydia Nahwooksy, former director of the Lincoln Indian Cultural Center; Dennis Hastings, Omaha Tribal Historian; and members of the Omaha Tribal Nation.

At the University of Nebraska–Lincoln, a number of colleagues were very supportive. Special thanks go to Dean John Peters of the College of Arts & Sciences for his sincere interest in my work. Many people at the Center for Great Plains Studies, including Lynn White, Frances Kaye, Linda Ratcliffe, Clare McKanna, Sharon Bays, Lisa Spaulding, Martha Kennedy, Elizabeth Mota, and Lori Gourama, provided encouragement in a number of ways. Sharon Bays deserves special notice for her help with the manuscript's preparation. Faculty in the history department, most notably fellow historians of the North American West—Frederick Luebke, Gary Moulton, and Ralph Vigil—plus former chair Benjamin Rader, and Jack Sosin, and Learthen Dorsey; and history graduate students Steve Potts, Lu Ke, James Hewitt, and Todd Kerstetter, have been extremely supportive. Other university faculty, David Wishart in geography and Elizabeth Grobsmith and James Gibson in anthropology; John Carter of the Nebraska State Historical Society; journalist Lawrence Goodman; and Mark Swetland and other students who teach and write about Native American studies and law have been very helpful and stimulating.

My debt of gratitude is greatest to Nell, Amanda, and Susan—our family. The benefits of receiving their knowledge and wisdom cannot be measured.

Lincoln, Neb.   J. R. W.
April 1993

Contents

"RETAINED BY THE PEOPLE"

1

Concepts of Rights

The history of Native American relationships with the federal Bill of Rights is an enigma, a perplexing state of affairs, for several reasons. First, the unique evolution of Native Americans in American law defies rational categorization. Historical periods from 1791 to 1991 are characterized by the frequent wholesale restructuring of that relationship. Although there is continuity in Native American legal history, it is more often marked by change.

Second, distinctions made between Indian individuals and Indian groups are extremely important when considering the relationship of Native Americans to the federal Bill of Rights. Definitions of rights are at cross-purposes and unsettled. Indeed, the entire concept of rights is at stake. Rights as defined in Native American society are limited, as few persons can hold a right against the world. This does not mean that individual rights are not protected and recognized, but consensus is sought to verify a right as one of the group's rights, not just one belonging to a particular individual.

Moreover, Indian legal relationships in the United States are perceived by Native Americans more as a struggle to obtain legal sanction and recognition of collective entitlements than as a denial of the enshrinement of individual rights. Cultural ignorance and intolerance centered in those from the dominant society enforcing the Bill of Rights often classify Indian disputes as obscure and inconsequential.[1] But nothing could be more incorrect, for at stake are cultural values and cultural survival.

A third aspect of this riddle is the strains of conflicting legal interpretation. Federal Indian jurisprudence is in constant conflict: It acknowledges limited forms of sovereignty within a forced union; it recognizes Indians as special groups that require special treatment in special circumstances; and it places Native American rights at the disposal of Congress rather than constitutional interpretation.[2] It is this last strain that denies the application of the Bill of Rights to Indians as Indians.

Walk down the street and ask the first ten people you encounter this one question: Does the Bill of Rights apply to everyone living in the United States today? After you recover from the shock that at least three of your interviewees are not familiar with the Bill of Rights (and so you select three more persons to interview who are knowledgeable), you will probably find that all ten of the random responses were yes.

Perhaps one or two of your new acquaintances might think that aliens were not completely covered by the Bill of Rights, but after further reflection they concluded that, yes, all are covered. You next have the task of enlightening those persons you surveyed. Tell them that not every person residing in the United States has been or currently is covered by the Bill of Rights and that even American citizens are denied these fundamental legal values. "Who might they be; what rights are in dispute; and why has this happened?" comes the amazed reply.

Culture and Conduct:
Considering the Dynamics of Native American
Individual and Collective Rights,
from Earliest Times to 1775

The first native peoples to occupy the land that became the United States arrived in their new world from Asia as long as 50,000 years ago. They came in small heterogeneous groups, spoke many different languages, were hunters and gatherers, and usually traveled by land. Smaller numbers of indigenous peoples journeyed by boat across the Pacific Ocean to the North and South American landmass.

Those formerly Asian peoples who came via the land bridge eventually migrated from what today is Alaska through Canada to the eastern slope of the Rocky Mountains. These migrations took thousands of years. Once in the foothills of the Rockies, native peoples had choices: They could stay in the region; they could cross the difficult terrain to the Pacific coast; they could travel over the Great Plains; or they could continue a southerly migration. All four alternatives were acceptable to at least some bands and tribes, but most chose the southern migration. When Europeans and Africans met these native peoples in their "Old World" after 1492, the major concentrations of indigenous peoples were in Mexico and Peru.

The physical evidence of these early peoples is sparse. The oldest artifact identified by archaeologists is a 27,000-year-old caribou bone scraper found at Old Crow Flats in Yukon Territory. The oldest skeletal remains have been dated as approximately 9,000 years old: Marmes man, found in Washington State, and Midland woman in West Texas. Nevertheless, by the fifteenth century, hundreds of thousands of Native Americans resided in what became the United States. During the sixteenth century, they met Europeans and Africans for the first time, and a culture clash of major proportions resulted from this encounter, including that of laws and legal systems.

Indigenous Concepts of Rights

Assessing native peoples' rights in pre-Columbian America is a difficult business. What evidence is available is largely hundredth-hand and passed down by oral tradition. Although there is a great deal to be said about the reliability of oral tradition, specific information is sometimes impossible to verify or has been shaped by a teller along the way to fit a particular situation. Overviews and generalizations, however, are very helpful. The problem of evidence also is complicated by the failure of non-Indians, when they first met native peoples, to ask the kinds of questions that would be useful to understanding indigenous concepts of rights. Both Indians and Europeans were highly ethnocentric and rarely bothered initially to find out about legal intricacies. The concept of rights did not seem to have entered their early discourse.

Subsequent fieldwork by many anthropologists in the late nineteenth and early twentieth centuries among Indians only recently forced to reside on reservations was not done much better. Legal anthropologists were more interested in debating whether there was such a thing as legal anthropology than in investigating the law of indigenous societies. Recent texts on political anthropology offer little enlightenment; they do not even define rights.[3]

Those few legalists and anthropologists who wrote about the law of indigenous societies did so mostly from a structural-functionalist approach or a process approach. Structural-functionalists assessed native legal systems as static and constantly seeking to adjust their societies to group consensus. Consequently, structural-functionalists did not address concepts of rights either inside or outside societies, and they were ahistorical. The process approach emphasized the use of public power by individuals and its impact on an indigenous society. The problem with those who advocated the process approach was that their studies were accultural and frequently overgeneralized the uses of power. They, too, were sometimes ahistorical, and their focus on political factions skewed the political and legal history of indigenous peoples.[4]

One anthropologist who pondered the *nature of law* and, to some degree, the concepts of rights in indigenous societies is E. Adamson Hoebel. Hoebel has much to offer to this discussion.[5] In *The Law of Primitive Man: A Study in Comparative Legal Dynamics,* Hoebel sees law as a part of the total fabric of culture. There are no clear-cut edges. Indigenous peoples throughout the world have social order and systems of social control. The elements essential to law, as Hoebel explains, are threefold: Law embodies the experience of the individual living within, not beyond, his or her society; law is dynamic, always changing while drawing on a consensus of the meaning of the past; and law offers society a means of predicting behavior.[6]

Hoebel criticized those who equate law and custom as one. That, to Hoebel, cannot be. Law reinforces custom or assists in making new custom.[7] For law to work, a legal system must contain a set of understood rules, predictable outcomes, decision-making courts, and enforcement mechanisms. The variables between native law-ways and nonnative legal systems are the courts and law enforcement. For native peoples, rights are found in rules and outcomes; for nonnatives, rights come from the courts upholding them and the institutions enforcing them against societal or individual wishes.[8]

Rights do exist in native peoples' societies. Hoebel defines indigenous concepts of rights as a system of dual rights: A *demand-right* leads to a corresponding duty; it is an active, positivist right. If a dispute is a demand-right, then resolving it will require the performance of a duty as a legal sanction. A *privilege-right* does not require a demand-right in return; thus a privilege-right is a passive, negative right.[9]

Fundamental differences existed, for example, between the Yuroks solving a problem and a colonial non-Yurok nation holding powers over the Yuroks deciding a legal issue. Yuroks at the time of European contact lived in southern Oregon and northern California along the Pacific coast. Yurok courts were composed of *crossers,* persons not related to those having a dispute. Thus clan and extended family relationships were extremely important to any dispute resolution.[10] The Crossers decided whether the laws of a tribe had been broken. These laws were enforced not by specialized personnel, but by members of the society who knew what to expect. For the Yuroks, restoring the social order included the collection of damages and even capital punishment. Everyone except slaves and bastards (because their mothers never received a bride-price) had legal status and legal rights.

In a case related by Alfred Kroeber, a noted cultural anthropologist, the headman of a Yurok family held the right to demand the flippers of all sea lions caught on a particular beach. But on several occasions, one Yurok hunter paid no attention to this right. After pondering this situation, the headman took matters into his own hands and wounded the father of the hunter with an arrow. The hunter's family then brought in the crossers to investigate and make restitution. The crossers decided that the damages of the wounding were less than the damages sustained by the headman and his family by the hunter's failure to provide the flippers. In other words, the demand-right required a duty that was more significant than the privilege-right not to be attacked in this particular circumstance. Supposedly the affair was settled, but as is often the case, the participants were not prepared to stop. The hunter was upset, and after the settlement, he cursed the headman. This necessitated further proceedings, and while they were being conducted, relatives of the headman killed the hunter. This was a terrible wrong that threatened the entire tribe, and so the crossers decided that the right to the sea-lion flippers should be transferred from the headman and his family to the mother of the dead hunter.[11]

On this occasion, Yurok law resolved a number of issues related to individual rights *within* Yurok society. At no time did the court allow individual rights to be asserted against other individuals or the tribe itself. The final solution was aimed at restoring the balance, bringing harmony to the nation, and making restitution.

What, then, are the *concepts of rights* for native peoples? Individual rights are those predictable rules reached by consensus within a society that are enforced by the society. They are held by the nation, the tribe, the band, the clan. Behavior, whether it be for the Senecas, the Cheyennes, or the Cherokees, was determined not by the actions of courts or law enforcement, but by duty, community predictability, and a fear of group ostracism or the unknown.[12]

Rights expressed beyond the indigenous nation are collective. To assert rights outside the native legal system is to attempt to enforce *collective entitlements*.[13] What is best for the group and what is required by the very nature of the tribe's existence form the dual nature of collective entitlements. Collective entitlements for Native Americans after 1492 must be seen within the notion of sovereignty.

Sovereignty is a human concept that has been defined in many ways, but two will suffice for us. Vine Deloria, Jr., describes sovereignty as evolutionary. For Deloria, the strongest sovereign nation has the ability to "determine its own course of action with respect to other nations," control "sufficient territory and military strength," and "regulate one's own internal functions in the field of domestic relations."[14] Under international law, sovereignty was defined most recently in Article I of the Montevideo Convention on the Rights and Duties of States. For a state to be a sovereign nation, it must have a permanent population, a defined territory,[15] a government that functions, and the ability to conduct relations with other states. According to these criteria, Native Americans have been sovereign nations for centuries and continue to be under international law.[16]

The particular kind of nation that one is depends on its ability to control its own destiny. Sovereignty is, however, only one component of complete independent political autonomy. Military strength and self-government also are fundamental to the assessment of a state's power.[17] Before 1492 until at least 1776, Native Americans lived in numerous states at various stages of political autonomy. Some nations, such as the Pequots in Connecticut, were barely able to exist, whereas others, such as the Haudenosaunee, or Iroquois Confederation—composed of the Senecas, Tuscaroras, Onondagas, Mohawks, Cayugas, and Oneidas— constituted formidable national powers in North America.

Some legalists downgrade the role of sovereignty in any consideration of Native American rights, suggesting that it can be lost in a variety of ways. Discovery and occupation, conquest, cession and agreement, and adverse possession or control over the land base for a long period are put forward as arguments to eliminate collective entitlements. But they have not generally held up, for as long as Native Americans have a permanent population that occupies a territory and has a legitimate government capable of representation, they are able to assert their sovereignty and reach for their rights. After all, as philosophers contend, rights exist whether or not they are guaranteed by a legal system, and ultimately rights mean consequences.[18]

The same legalists have not come to grips with a fundamental concept: Indian tribes are nations fully capable of making treaties,[19] deciding what is in their national interests, operating legal systems within their nation, and ascertaining their collective and individual rights. Such legalists see sovereignty in absolutist terms, but as Charles Wilkinson points out, the establishment of the United States and its unique form of constitutional government resulted in a system of diffused sovereignty, a sovereignty of dual federalism. Since the creation of the United States Constitution and the addition of the Bill of Rights to the Constitution, sovereignty for Indians when considered within the legal system of the United States has no realistic relationship to absolutist notions. Instead, constitutional applications to Native American rights have been fluid and controversial.[20] As Vine Deloria, Jr., and Clifford Lytle explain, "American Indians are unique in the world in that they represent the only aboriginal peoples still practicing a form of self-government in the midst of a wholly new and modern civilization that has been transported to their lands."[21]

Native Americans in 1491

By the time Europeans and Africans arrived in North America, Native Americans had called it home for over 20,000 years. Theirs was a dynamic existence punctuated by many changes in a multitude of native societies. At least three phases have been identified by archaeologists.

The first is the Pleistocene Age, a stage lasting from approximately 25000 B.C. to 8000 B.C. During this time, technologies improved so that native peoples termed *Paleo-Indians* depended less on stones than on shaped knives and projectile points. Hunting societies were more proficient, and material goods proliferated. The next phase, the Archaic Age, spanned nearly seven centuries, from 8000 B.C. to 1200 B.C. During this era, major environmental changes occurred that hastened alterations in Paleo-Indian economies and diffused their settlement.[22]

Eventually major civilizations emerged during the third stage, the Age of Early Modern Tribalism, after 1200 B.C. In the Southwest, the Anasazi, Hohokam, Mogollon, and Patayan adapted to semiarid and arid lands by domesticating plants and animals and developing sophisticated irrigation works. Massive structures were built, and large concentrations of sedentary peoples converged in urban settings. In the East, what has been termed the Mississippian Culture arose around A.D. 500 and lasted until 1250. Huge cities with extensive trade networks dominated the interior of the continent. Both the southwestern and eastern civilizations perished. Why they did is subject to debate; a number of reasons have been posited that range from catastrophic environmental changes, plague, and internal revolutions, to invasions of new nomadic peoples and despotism.

By 1491, there were many Native American nations throughout North America. Population estimates range from several million to hundreds of millions. In terms of language alone, at least 170 separate languages were spoken by tribes located north of Mexico. Differences also characterized the tribes' economic, social, and political organizations. Hunting, farming, herding, gathering, and limited manufacturing societies existed side by side. Matrilineal and patrilineal societies built close-knit kinship linkages. Indian religions were unique to each tribe or band unit.

Politically and legally, many systems were employed, such as patriarchies, chiefdoms, theocracies, and matriarchies. Clan systems provided village governments with loose intervillage cooperation, and clan leaders and a council of elders set policy and law. Both private property and property held in common existed. Courts, enforcement mechanisms, and a system of private and public law were in place. Confederacies emerged in several regions, where they acted primarily in foreign policy matters, although there were efforts to centralize local law.

The year before Columbus stumbled his way into the "Old World" of the native peoples of North and South America, the geographical demarcation of Indian nations covered the entire expanse of the continents. The Northeast woodlands and the Great Lakes region featured Algonkian, Iroquoian, and Muskhogean speakers. Among the Algonkians, some, like the Passamaquoddys of Maine and the Menominees of Wisconsin, were seminomadic hunters and fishers who lived in loose-band confederacies. Others, like the Wampanoags of Massachusetts and the Shawnees of Ohio, were farmers who constructed military confederations. The Haudenosaunee, or the League of Five (later Six) Nations, represented an Iroquoian confederation of agriculturalists who developed a complex matrilineally dominated legislative and legal system. In the Mississippi and Ohio valleys, Siouan-speaking tribes were moving west, in part because of the expanding Algonkian and Iroquoian nations. Omahas, Osage, and Yankton and Oglala Sioux, for example, practiced a mixed economy and were beginning to change their kinship systems from a matrilineal to a patrilineal base.

In the South, the Cherokees lived in a series of villages near the Appalachian Mountains. These Iroquoian speakers embraced a clan law system of some intricacy. Near the Cherokees were a number of Muskhogean-speaking tribes, including the Creek, Choctaw, and Chickasaw, who also had developed a clan-based legal system in their independent city-states. They were agriculturalists who also hunted and gathered to supplement their economies. The Natchez, a large tribe located near the Mississippi River, retained a complex theocratic political and legal system.

The region west of the Mississippi also provided the sites for numerous nations. In the Great Plains, a region of significant tribal movement, a number of tribes—including the Kiowa, Comanche, and Crow—lived on the western margins in 1491. They were nomadic hunters of the bison who ventured onto the Plains and returned to the foothills of the Rockies. In the center of the Plains were the numerous divisions of the nomadic Apache hunters. Although they had been on the Plains for several centuries, they were gradually losing their hegemony to the tribes on the western and eastern margins. The Apaches and the other Plains nomadic hunters lived in small bands that decided questions of law within their societies through kinship and consensus. In the valleys of the Plains were agriculturalists such as the Mandans in the north and the Pawnees in the central Plains. All the Plains agriculturalists lived in semipermanent settlements, sometimes numbering several thousand. Here village councils based on kinship decided political and legal policies.

In the Southwest resided a number of tribes with long-standing traditions reaching back to the Archaic Age and the Anasazi and Hohokom. They included the Pueblos and the Hopis, who had traditional councils and legal system specialists to decide individual disputes. They were agriculturalists confronted by nomadic Athapascan speakers, the

Apaches and the Navajos. The Athapascans lived in autonomous bands with no village or central government.

Moving north into the desert areas of the Great Basin were the Shoshonean peoples, nomadic hunters and gatherers who lived in patriarchal band organizations. Farther north on the Columbian Plateau were fishers and hunters, some nomadic, such as the Nez Perce, whose lifeways resembled those of the nomadic hunters of the northern Plains. The Pacific Northwest included the Salish peoples, who were hunters, gatherers, and fishers. They lived in small villages and built a legal system around the potlatch. Similarly, in California, natural food resources allowed for a vast population that congregated in village societies. These hunters and gatherers were the most diverse of the cultures in North America, speaking more language variations than any other peoples in the world.

On the coasts of Alaska were Pacific Northwest tribes such as the Tlingit and Haida, who hunted, fished, and lived in independent village states. In the interior were Athapascan-speaking seminomadic hunters. Like their cousins of the Plains and Southwest, they conducted political and legal affairs based on kinship and band relationships.[23]

Thus in 1491, hundreds of tribes in a tremendous cultural mix with many different legal systems covered North America. These Indian nations, according to what can be learned from oral traditions and the limited existing scholarship, perceived of rights in two ways: Individual rights were a matter for internal dispute resolution that took into consideration the best interests of the tribe, band, or village; kinship ties; and a number of balancing or harmonious outcomes. Group rights, or collective entitlements, concerned external relationships with other villages or other nations. They were a matter for negotiation and diplomacy, confrontation and blood revenge. North American indigenous peoples who held these generalized concepts of rights soon were confronted by non-Indians and their very different concepts of rights. No accommodation was possible. How native individual and collective rights emerged after and during their struggle against an external colonial power is the focus of this inquiry into the relationship of Indians and the Bill of Rights.

The Colonial Era

After 1491, the Christian nation-states of Europe invaded the Americas. Of significance to Native Americans north of Mexico were their encounters with fur traders, settlers, farmers, soldiers, clergy, and legalists from Spain, France, England, Holland, Sweden, and Russia. The rights of America's indigenous peoples were the subject of great debate among European philosophers and jurists. Central to the discourse were the native inhabitants' sovereign rights to their land. The prevailing sentiment established that Native Americans could not own the lands they occupied and that genocide was justified because Indians were not humans. Real property could be confiscated because Native Americans as non-Christians did not use their land and were simply beyond all natural law.

Europe's conquest obtained sanction from its legal theories. Put in practice, such theories could be startlingly cruel: Spain readily turned to the *encomienda,* a system of forced labor and slavery in which thousands of Indians perished. England used wars and licensing techniques to deny the existence of native rights. Wars were justified because of the doctrine of discovery and the belief among all the European colonizers that they had a right to wage war against infidels. Treaties that often were the end product of wars were devices used to justify land grabs. Their legal use supplemented rather than displaced the greater doctrines. In addition, advanced parties of English settlers regulated trade with native peoples through licensing and contracts that accorded few rights to the Indian participants.[24]

A particularly noteworthy crisis came in seventeenth-century Virginia when legal rights within the English framework did not match the settlement's military might. The swamp-ridden Virginia Company outpost was not powerful; its military strength was less than that of the surrounding twenty-eight-nation Powhatan Confederacy. Although company officials tried to persuade the confederacy's leader, Powhatan, to defer to them, they could not force him; after his death, in order to maintain trade relationships with the Indian confederacy, the company found itself agreeing to obtain the permission of Powhatan's brother Opechancanough, the Indian confederacy's new nationalist leader, before any land title could be conveyed. A Mr. Barkham submitted his title for confirmation to the Virginia Company's London-based council, which rejected it because it contained language of joint permission—from the company and from Opechancanough.[25] Legal doctrine thus was more important than reality. This kind of legal fiction haunted the English during the American Revolution.

In general, disruption and confusion marked the colonial period.

Spain situated missions and presidios on the borderlands of what would become the United States—in the Southwest, Texas, and Florida. England, through superior population importation, established a beachhead along the Atlantic coast. Limited attempts at colonization by Sweden and Holland were squelched by the English. France moved down the St. Lawrence River basin into southern Canada, the Great Lakes, and the Mississippi Valley. Short-term resource exploitation, that of the fur trade, limited the eventual empire amassed by the French. Similarly, Russia toyed with settlements in Alaska and as far south as northern California.[26]

Eventually England prevailed, and it is against its legal traditions that Indian concepts of rights abutted. Ironically, part of the challenge from the English colonists to the Crown during the American Revolution was over the right of Native Americans to be sovereign. Settlers wanted to be able to obtain with a clear title lands from native peoples. But the Crown did not wish to relinquish this power, and moreover, it ran counter to the established legal thought prevailing in Europe since medieval times. There, of course, were other grievances, but this one is especially relevant to the issue here at hand.

To curb colonial "irrationality," the Crown set about to restrain contractual trade relationships and to prevent frontier real-estate development. After defeating the French, England issued the Proclamation of 1763, which prohibited settlement west of the Appalachian Mountains, restricted trade, and placed the colonists on notice that the Crown would not tolerate Indian real-estate "deals." The Stamp Act further infuriated the colonists not only because of the "freedoms" they lost, but also because this was further evidence of the king's forcing them to pay for his war, the end product of which was to deny Indian–colonial contractual relationships. But some of the more radical colonists believed that the Native Americans were indeed human and had natural law rights similar to their own. To them, treaties were contractual agreements between nations that had the force of law.[27]

The American Revolution came and went, but it did not result in any significant change as far as the rights of Native Americans were concerned. Those living in the new United States signed treaties, became the subject of trade agreements, and witnessed the disorder of a new nation moving from a confederation to a centralized state. Indian collective entitlements depended on force, not on legal theory developed during these formative years. Once it became clear that a new nation with a new constitution would be applying that document to its colonized native inhabitants, the issue of rights came to the fore. Indian rights would be the subject of over two centuries of continual debate, suppression, confusion, and denial.

■   ■   ■

Hundreds of years ago, the Pueblo peoples congregated in the Southwest. There they built a strong civilization that lasted after the European invasion of the Americas. One of their own, Simon J. Ortiz, recently described in verse the first meeting of his people with Europeans. The Europeans offered Christianity to the Pueblos, but Ortiz's people "knew the sacredness of life, knew that we had to keep the belief and practice of regeneration always within ourselves, and we knew that this was a way of life that had been learned from many years of living."[28] But the Europeans would not leave. They were "wildeyed, frantic with hunger and thirst." After food and water did not satisfy them, what did they want? "For them, it was what they said: the power and will is with us and we want you [the Pueblos] to submit and be humbled. It wasn't an issue of religious belief to them; they wanted power over the People, yes, and over the land."[29]

Colonial power over Indian peoples and their lands constituted a direct destruction of individual rights and collective entitlements. Might a bill of rights have prevented the loss of rights for Native Americans?

2

The Old and New Colonialisms

From the constitutional beginnings of the United States, the American nation acted as a colonial power. Such is often the nature of new nations. It successfully broke away from one of the world's superpowers and, in the process, set up a democratic-governing experiment. Here was a weak republic in the world of kings and queens. Even though the theoretical basis for revolution had been the oppressive actions of King George III, who had denied the English colonists those freedoms held dear by other Britons, the United States' anticolonial revolutionary fervor could not be maintained. The colonial past of the new American republic was bound to be repeated: The pupil learned from the teacher how best to subjugate neighboring peoples. Native Americans soon found that the new republic was committed to the same land-expansionist tendencies and sovereignty restrictions that characterized the Europeans of the previous century. Rights as collective entitlements or individual liberties provided minimal resistance in a world predicated on force rather than law.

Colonialism has been defined in many ways. Indians recognized the concept in their initial relationships with the United States. Ben Kindle's Winter Count for the year 1791 expressed the early sentiment of the Oglala Sioux: "We' mapi mak'o' 'Kawih ahi'yayapi'" (Flag around the earth / they carry it along).[1] Before Ben Kindle drew his Winter Count, Dragging Canoe, a Cherokee, observed, "We had hoped the white man would not be willing to travel beyond the mountains [the

Appalachians]; now that hope is gone.'' In 1768 at one of the councils involving the Cherokees and the British, Dragging Canoe predicted,

> Finally, the whole country, which the Cherokees and their fathers have so long occupied, will be demanded, and the remnant of the Ani-Yunwiya, ''The Real People,'' once so great and formidable, will be obliged to seek refuge in some distant wilderness until they again behold the advancing banners of the same greedy host.[2]

Dragging Canoe's characterization of the United States as a future ''greedy host'' came true. Americans practiced a simplistic brand of colonialism that was wedded to a treaty system institutionalized before 1871, the year that all treaty making between the United States and Native Americans ended. After 1871, a new form of colonialism emerged. This *Old Colonialism,* the same colonialism that Europeans followed in Africa, North and South America, and Asia, had as its primary goal the physical acquisition of valuable western and southern lands and the physical subjugation of its peoples. Native peoples' rights or sovereignty, including any legal traditions, deferred to the will of the conqueror.

By the 1870s, the Old Colonialism had run its course. The United States had achieved most of its territorial limits, had survived a fractious civil war, and was experiencing new settler demands for lands protected by existing treaties. The treaty system was in the way, and so a *New Colonialism* evolved. It was an especially virulent strain, gathering its strength and embellishment from legal argument and pronouncement. It attacked every aspect of Native American life—religion, speech, political freedoms, economic liberty, and cultural diversity. The United States Congress and the United States judiciary reinterpreted international law and congressional statutes in order to impose a stringent new legal regimen of colonialism on Native Americans, particularly those on reservations. Indian rights were severely restricted.

United States colonialism separated Indians from their lands by using a variety of methods. Confrontations and wars provided some means by which native peoples were dislodged from their homelands. On other occasions, the presence of a superior military force backed by significant population advances was sufficient to cause migrations. But in most cases, some sort of contractual agreement was made among the national parties seeking the lands. The presence of earlier inhabitants was recognized if only because law and tradition required an agreement, a *treaty*. The treaty thus became the vehicle by which the "greedy host" expanded the colonial empire. The treaty also was an instrument of law that cooled national tensions, demanded a sense of fair play and a meeting of minds, and was explicitly recognized by all nations, European and Native American alike, as a basic document inherent in any legal relationship. Treaties recognized sovereignty, so those who wanted to eclipse sovereignty attacked the basis of the treaty—the land and the culture of the people.

Reprisal and Remonstrance: Attempting to Destroy Nations Within a Nation, 1776–1903

After the American Revolution, Native Americans under United States law fit within a familiar colonial framework. The first attempts to place Native Americans under the Constitution and the Bill of Rights did not satisfy federal officials, so instead they used trade legislation to define the rights of Indians. Such definitions, however, amounted to assaults on various fundamental aspects of rights, and Native Americans resisted.

As the new American nation gained strength, both the settlers and the federal government sought to remove Indians from east of the Mississippi River. Indian removal, however, was more than simply a move. It included dislocation and physical and cultural genocide. So outrageous was the treatment of Indians that the Supreme Court attempted to intervene. Reprisals were swift, and a new legal colonialism was born.

Native Americans Within an Initial Constitutional Framework

Indian relationships with the United States did not begin in a vacuum; there were ample precedents to be gleaned from Dutch, French, Spanish, and, mainly, British experiences. But the colonial construct and preexisting treaty system were interrupted by the American Revolution, and the political and legal culmination of the Revolution was centered in the Articles of Confederation. After being adopted in 1781, the Articles were quickly found wanting by American leaders. They seemed too loose to many and lacked the centralizing power necessary to retain the independence gained by the Revolution. The conduct of foreign relations was especially confused, and nowhere was this confusion more obvious than in Native American–United States relationships.

Already agreements had been negotiated that needed to be honored. The first treaty between the United States and an Indian nation, the Delawares, was signed in 1778. So concerned was the young Congress about the delicate diplomatic balance during the Revolution that the Delawares were offered the possible opportunity of statehood,[3] but this offer was never renewed after the Revolution was over. By 1790, virtually all the tribes residing in American states on the Atlantic coast had signed agreements with the United States. These agreements restricted American settlers from taking Indian lands, but they also restricted Native Americans from making alliances with or ceding land to other nations. So diffuse was power in the new nation that many states also concluded treaties with tribes in or near their state boundaries. In the twentieth century, these agreements became important legal documents for the Native Americans of New England and the Southeast.

Any rule of law under the Articles of Confederation became futile because of actions by the states and their frontier settlers' attitudes toward Indians. Agreements signed by the state and national governments were ignored, and confrontations arose throughout the frontier areas. The treaty system, so much a part of the colonial regimes of the past, needed a centralized focus in order for a national government to survive. Of course, other destabilizing factors were involved, but confusion over the treaty system was chief among them.

When delegates met at Philadelphia in 1787, they discussed at length power in conjunction with the enlightenment goals of life, liberty, and the pursuit of happiness left over from the Revolution. Recent scholarship argues that the Iroquois Confederation experience was on the minds of a number of delegates at Philadelphia, that several Constitutional Convention members who were in a position to shape the writing of the Constitution had observed the Six Nations republic.[4] Whatever amount of Iroquoian political theory was incorporated into the Constitution, the resulting document said little specifically about Native Americans.

Indians are mentioned directly only two times in the Constitution. Article I, Section 2, provides that "Representatives and direct Taxes shall be apportioned among the several States which may be included within this Union, according to their respective Numbers . . . excluding Indians not taxed."[5] Section 8 of the same article gives to

Congress the power "to regulate Commerce with foreign Nations, and among the Several States, and with the Indian Tribes."[6]

The delegates took Native Americans into account in two other ways. Again in Article I, they stipulated that "no State shall enter into any Treaty Alliance or Confederation" or "enter into any Agreement or Compact with another State, or with a foreign Power."[7] The lessons of the Articles of Confederation would not be lost. But treaties between the United States or the states and Native Americans made before 1787 would be honored because of Article VI. Here the Constitution stated bluntly:

> This Constitution, and the Laws of the United States which shall be made in Pursuance thereof: and all Treaties made, or which shall be made, under the Authority of the United States, shall be the supreme Law of the Land; and the Judges in every State shall be bound thereby, any Thing in the Constitution or laws of any State to the Contrary notwithstanding.[8]

Taken as a whole, the Constitution emphasized order and stability. Gone were the heady days of the Declaration of Independence when Enlightenment ideology embraced liberty and freedom. By 1787, events seemed to reflect a need for checks and balances in government, centralized power, and majority rule. But what about minority rights?

Enough Americans were concerned about this apparent deficiency in the new document to require a bill of rights. The first ten amendments, ratified in 1791, embodied a number of personal freedoms that both Indian society and that of the new United States considered sacred or basic to life itself. In some ways, the Bill of Rights represents a code of criminal law, but it is more complex than this. Five of the amendments specifically concern the "rights of the people." Five others guarantee basic rights of criminal procedure against governmental oppression.

The amendments that make up the Bill of Rights were broadly defined, and through the years they have been interpreted in a variety of ways that have expanded their application. But what is important here is whether the Bill of Rights amendments applied to Native Americans in 1791. What was meant by the phrase "of the people" in these amendments? The preamble of the Constitution begins with "We the people of the United States . . . ," suggesting that the people must be residents of the United States in order for them to be covered by the Constitution. Might Native Americans be covered once they signed treaties turning over much of their lands to the United States? Did the Constitution include Native Americans residing on reservations within the geopolitical boundaries of the United States?

From the very first applications of the Bill of Rights, Native Americans have not been covered by its penumbra. Legislators and judges stipulated that Indians were unique in many legal ways. Although they were not yet citizens of the United States, they also were not aliens. Congress and the Supreme Court therefore moved to develop laws that created a new kind of legal life for Indians within the American legal system. Indians, after all, held sovereignty over the lands that became the United States, and this allowed for a residue of sovereignty left over in the tribes after the signing of treaty documents. If an Indian ceased being an Indian per se and renounced his or her tribal status, then the Bill of Rights might apply. Thus in their relationship with the first ten amendments, Native Americans were considered both pre–Bill of Rights and extra–Bill of Rights.

Trade Laws and Impingements on the Bill of Rights

Of primary concern to officials of the United States during its first half-century were trade relationships with Native Americans. Congress passed a series of commercial statutes that were intended to be used to energize the treaty system. The very nature of these laws put great restrictions on native peoples, with infringements placed at the heart of the initial federal policy toward Indians. As citizens of a budding colonial power, Americans sought to maximize their international opportunities, such as they were for a small, principally agricultural republic. Commercial ties to Indian tribes helped the United States gain greater access to frontier markets and increased opportunities to extend federal control over Indian lands. Trade and colonialism marched hand in hand.

In 1790, Congress passed the first law governing commerce between the Indian tribes and the United States. This was a broadly based attempt at the regulation of Indian country by the United States. Entitled ''An Act to regulate trade and intercourse with the Indian tribes,'' the act restricted all trade with Indians to those traders who received from the federal government a license to trade. The act also set standards for traders to meet in order to retain their licenses, required the sale of Indian lands only to the United States government, established the notion that crimes committed by non-Indians against Indians should be tried in American courts, and instituted a procedure for punishing non-Indians found guilty of committing a crime or trespassing on Indian lands without the express authorization of the United States government.[9]

This first attempt at trade legislation was much more than simply sovereign powers offering most-favored-nation status to one another. From the United States government's legal perspective, the commerce clause of the Constitution was being implemented. But in addition those rights to be covered in the Bill of Rights, adopted the very next year, were being subsumed. Fundamental forms of criminal procedure were co-opted from Native Americans for the federal courts, and property sale restrictions at odds with the future Fifth Amendment were put in place.

Subsequent trade legislation passed by Congress placed even greater restrictions on Native Americans than did the initial Trade and Intercourse Act of 1790. Indians accused of committing crimes off their reservation could be prosecuted only in federal courts. Settlement was restricted on Indian lands, an action that also diminished sovereignty. Passports were required of non-Indians to travel in Indian country, and the government instituted its own monopoly over trade by creating officially sanctioned trading houses or factories.[10]

Perhaps the greatest attack on Indian rights came in 1800 with the passage of "An Act for the preservation of peace with the Indian tribes."[11] Native Americans were prohibited from discussing diplomatic options, and they were not to criticize the United States government. The rights of free speech and free press could be curtailed in American jurisdictions. This trade regulation no doubt was aimed at preventing Europeans from influencing various Indian tribes, and it did come at a time of anti-French frenzy in the United States that resulted in the Alien and Sedition Acts of 1798, but the language was explicit. Trade regulation for Native Americans had been expanded to suppress rights found in the First Amendment, and it was thirty years before tribes were able to obtain their first Supreme Court opinion on their rights.

Several months later in 1800, Congress passed another trade provision that had profound ramifications. This law authorized the passing out of rations at army forts to Indians, and it sought to restrict the flow of alcohol to Native Americans. This form of tribute was used to encourage positive commercial relations and to anchor a colonial system of the future, in which the United States enforced a specific policy effort to make Indian tribes dependent on the federal government for assistance.[12] Dependency, then, facilitated the erosion of sovereignty and encouraged the postponement of any application of the Bill of Rights to individual Native Americans wishing to retain their tribal affiliation and treaty rights.

During the 1810s and 1820s, Congress continued to link Indian rights to a trade policy. New legislation included provisions for extending criminal justice to Indian country and the first of many attempts by the federal government to alter Native American cultures. Indians committing a crime on their reservation against non-Indians now could expect to be tried in a federal court, even though Bill of Rights guarantees concerning criminal procedure were not yet applicable to Indians.[13] In "An Act making provision for the civilization of the Indian tribes adjoining the frontier settlements," Congress specifically made it United States policy to offer Native Americans agricultural training; strong role models through moral Indian agents; and reading, writing, and arithmetic instruction for Indian children. Such "benefits" were to accrue only with the consent of the Indian tribes.[14]

Trade legislation culminated in 1834 with the passage of the all-encompassing Trade and Intercourse Act. This landmark legislation was the result of constant requests by Indian agents for a simplified code and the result of the first of many official reports from the House of Representatives Committee on Indian Affairs complaining that Indian relationships were in need of "reform" in order to "civilize" Native Americans. It was drafted by territorial governors Lewis Cass and William Clark and included some fifty-six sections with commentary justifying each proposal. Most of the sections were restatements of old statutes; few changes were incorporated into the comprehensive trade regulations package. The only major alteration relaxed the requirement that all persons needed a passport to enter Indian country. Under the 1834 act, only foreigners needed a passport.[15]

Thus federal policy toward Native Americans was firmly anchored by 1834. The treaty system provided a useful vehicle by which trade regulations could be used to deny fundamental forms of fairness to Indians. Passed ostensibly for the good of various tribes, these statutes, culminating in the Trade and Intercourse Act of 1834, amounted to a taking of basic rights and an infringement on tribal sovereignty. The passage of these trade laws and their subsequent implementation meant that the Bill of Rights probably would not be applied to Indians in the United States.

Indian Removal and Its Legal Challenge

An even more ominous threat to the rights of Native Americans than trade legislation came in the form of the federal policy of *Indian removal*. In May 1830, Congress passed the first national Indian removal act.[16] Previous legislation at the national level had been piecemeal, primarily confined to treaty negotiations, or states and localities had taken matters into their own hands.

The removal law solidified as national policy an exchange-of-land concept. Indians living east of the Mississippi River signed documents accepting new homelands in the Louisiana Purchase lands west of the Mississippi River or at least in sparsely settled regions bordering the Mississippi River. Thomas Jefferson first advanced this idea, ostensibly to assist Native Americans to become acculturated with minimal disruption. Andrew Jackson, however, implemented the plan, and his motivation contained few humanitarian notions.

Portions of the law appeared to protect Indians, as it pledged that the United States and the president guaranteed to all those tribes that were removed a secure patent to their new lands. Moreover, any signed agreements would be based on free will. Such a voluntary agreement would constitute a contract made without coercion. In addition, Congress granted the president summary powers over the removed tribes. The President was required to "cause such tribe or nation [removed] to be protected, at their new residence against all interruption or disturbance from any other tribe or nation of Indians, or from any other person or persons whatever."[17] Of import was a last caveat: "that nothing in this act contained shall be construed as authorizing or directing the violation of any existing treaty between the United States and any of the Indian tribes."[18] All treaty rights were to be protected.

Removal in theory and removal in practice were worlds apart. By authorizing the federal bureaucracy to begin entering into unsupervised and unregulated local negotiations with eastern tribes to take their lands, Congress willingly opened the door to fraud, coercion, deceit, and violence. Once the gross mismanagement and fraudulent activities of frontier entrepreneurs were made known, the United States Army was sent in to supervise the exodus.[19] Perhaps Alexis de Tocqueville, a French visitor to the United States at the time of removal, was the most accurate of the contemporary observers who commented on the relocation of eastern Native Americans. He described the actual act of removal as essentially a friendly death march, writing that "the most grasping nation on the globe" was prepared to implement a policy of genocide "with singular felicity; tranquility, legally, philanthropically, without shedding blood." Tocqueville concluded that "it is impossible to destroy man with more respect for the laws of humanity."[20]

The Indian tribes' response to removal varied. Many, after initially refusing to sign documents, succumbed to the tactics, agreed to a removal treaty, and eventually moved west. Others resisted; for example, the Sac and Fox refused to move, and so the Illinois militia and United States Army were enlisted to compel their removal. Native Americans also tried to stop removal through legal channels.

The Cherokees' response was to take these matters to the American court system, where they challenged the constitutionality of the concept of Indian removal. But they embarked on their legal journey with impediments. Earlier, in 1823, Chief Justice John Marshall, using a fact pattern concerning Indians and their title to frontier lands, had constructed a preliminary foundation for placing Indian tribes under the American legal system. In *Johnson v. M'Intosh* (1823), Marshall held that Indian tribes possessed ownership of their homelands, including the rights to occupy, hunt, and fish on that domain.[21] However, according to the chief justice, the United States also shared title to the same lands, and consequently tribes could sell their land only to the United States. Similarly, the United States could not sell the land to any other nation or citizen unless it held a clear title, and this required the federal government to negotiate with the affected tribes, through treaties, any cessions of land.

Eight years after *Johnson v. M'Intosh,* the Cherokees challenged the state of Georgia's attempts to assert sovereignty over their lands. Georgia adopted laws that organized the Cherokee lands into counties and made Georgia law controlling. Georgia did not enforce these laws at first; it waited to see what the national administration of Andrew Jackson would do, and he did not disappoint the Georgians. Jackson embraced the state's position and told the Cherokees that they could either move west or stay in Georgia under state law.

The Cherokees were not prepared to accept removal without resistance; and so they hired a well-known Washington, D.C., attorney, William Wirt, to challenge the Georgia law. Meanwhile, Georgia officials decided to act. They arrested a Cherokee, George Tassel, who was convicted of a murder in Cherokee country. Although the conviction had been appealed to the United States Supreme Court, Georgia went ahead and executed Tassel before a decision could be made. On January 1, 1831, the Cherokees served notice on Georgia officials, requiring a hearing before the Supreme Court. Georgia refused to answer. Wirt then appeared for the Cherokees, and he requested an injunction to prevent Georgia from further actions. He argued that Georgia had violated Cherokee sovereignty, their right to self-government, and Cherokee–United States treaties.

Although the Supreme Court justices were sympathetic to the Cherokees, they found in *Cherokee Nation v. State of Georgia* (1831) for Georgia and refused to assume jurisdiction.[22] In order for the Court to consider the matter, Marshall said that the Cherokees had to be a "foreign state." He concluded that the Cherokees and other Native Americans shared a unique legal relationship with the United States, summarized in three words: The Cherokees were a *domestic, dependent nation*.

Marshall interpreted the Constitution to mean that Indian nations were not foreign nations, in part because they were classified separately in the commerce clause, and he described individual Indians as being in a state of pupilage; that is, Native Americans were the wards of their guardian, the United States. Thus as an entity, Indian nations could not possibly be foreign nations, and as individuals, Native Americans were indeed dependent and domestic peoples with some aspects of sovereignty retained through their treaties. Although Marshall never touched on the merits of this case, he did not seem comfortable with the results.

Justice Smith Thompson vigorously dissented, arguing that the Supreme Court did indeed have jurisdiction. He set up a model to determine what constituted a sovereign nation and concluded that the Cherokees were sovereign, albeit weak. Thompson noted that the Cherokees were trying to model themselves after the United States, and so to attack this development was to strike at the fundamentals of American republicanism, the Constitution, and the Bill of Rights. Marshall was more than listening.

One year later, Marshall forged a new consensus in another Cherokee case. This time Samuel Worcester, a white Congregational missionary and postmaster at New Echota, capital of the Cherokee Nation, was arrested by Georgia officials and charged with violating a new Georgia statute that prohibited white persons from residing in Cherokee Nation lands. To prevent Worcester from claiming that he was a federal employee, the Georgia governor persuaded President Andrew Jackson to fire him. Worcester was then rearrested, recharged, convicted, and sentenced to four years in prison. Worcester, a citizen of Vermont, appealed Georgia's actions to the Supreme Court.

In *Worcester v. Georgia* (1832), Marshall heard Wirt argue once again that the Georgia laws were void because they violated Cherokee–United States treaties, the Constitution's contract and commerce clauses, and Cherokee sovereignty.[23] This time, Marshall found that the state statutes were not applicable. Marshall essentially adopted Thompson's dissent in *Cherokee Nation v. State of Georgia*. Marshall found the Indian nations to be "distinct, independent political communities, retaining their original natural rights." Even though they were made up of a "people distinct from others," the Court considered them like "other nations of the earth."[24]

In a curious way, Marshall's three opinions represent a legal trilogy forming the basis for Native American existence within the American union. Indians were to be considered domestic, dependent nations, but they retained a distinct political nature as well as certain natural rights, rights guaranteed to other Americans by the Bill of Rights, and a tribal immunity from state law.

What happened to the *Worcester* case? Because of its timing, the Supreme Court's need for specific papers, Andrew Jackson's refusal to cooperate, and Georgia's intransigence, Worcester was not released. He eventually served out his sentence, and the Cherokees were forced to move to Oklahoma on the genocidal Trail of Tears, along which it has been estimated that as many as 10,000 Cherokees perished. Both *Worcester* and *Cherokee Nation* left a legacy on which the courts later relied heavily; indeed, the *Worcester* decision remains one of the cases most often cited in American constitutional law.[25] Whereas nineteenth-century law focused on defining "domestic" and "dependent" for nineteenth-century audiences anxious to take Indian lands, twentieth-century courts after World War II expanded the meaning of "nation."[26]

Prelude to the New Colonialism

The period from the Cherokee cases of the 1830s to the end of treaty making in 1871 represents the waning of the Old Colonialism. Its hallmarks—American expansion and the removal of Native Americans from their homelands—continued, but its certainty was clouded. Residual sovereignty rights were allowed to remain in Indian tribes, thereby complicating attempts by many Americans to obtain Indian land.

The federal government to some extent recognized this changing process when the Bureau of Indian Affairs (BIA), created in 1824, was moved from the Department of War to the Department of the Interior in 1849. There were bureaucratic and political reasons for the change in jurisdiction, but it reflected a new attitude as well. This transfer symbolized the new dimensions of the legal relationship between Native Americans and the United States. What previously had been an exercise in diplomacy or war, in which the major participants used by both sides were treaty negotiators, ambassadors, and warriors, now became attached to a small but growing government bureaucracy whose members were interested mainly in social and economic management.

But before new policies could be tried, the American republic burst at the seams. In the 1850s, the United States divided, and in 1861 the Civil War began. During this crisis, the federal government ignored more often than not its relations with Native Americans. Thus many Indian nations were left to try to sort out how best they might handle new diplomatic initiatives. American settlers also wanted to establish their claims to Indian homelands, and without the occasional restraining posture of the United States government, confrontations became commonplace. Sioux Indians were driven from Minnesota in 1862, and peaceful Cheyennes and Arapahos were brutally massacred at Sand Creek in southeastern Colorado in 1864.

In the confusion, state governments sought to make inroads into Indian rights, particularly by extending their taxation powers. The Kansas legislature passed laws taxing tribal lands and treaty allotments. Shawnees residing in Kansas challenged the legislation, and in 1866 the United States Supreme Court spoke to this issue in *The Kansas Indians* decision, ruling that even though the Shawnees were in a state of cultural transition, they were still "a people distinct from others." Kansas could not tax Native Americans because they were governed exclusively by Congress. This was guaranteed by treaties, the Constitution, and statutes of Congress itself.[27]

The Kansas Indians case, however, did encourage those who wanted to force Indians off their homelands and eliminate their cultural distinctiveness. The Court found that Kansas and other states could intervene if they were able to persuade Indian tribes to terminate their tribal organization or if Congress conferred national citizenship on Native

Americans. As long as an Indian tribe maintained its national character, the treaties applied, but if somehow tribal recognition ended or citizenship was bestowed, the treaties would cease to function.

After the Civil War ended, the federal government assessed its frontier situation and found its relations with Indians in a shambles. The Board of Indian Commissioners was established, and an Indian Peace Commission issued a report finding that the treaties were to blame for many of the problems. By attacking these treaties, the basic framework of the Old Colonialism came tumbling down.

The Resolution of 1871

Many Americans, for a variety of reasons and motives, concluded that the treaty system had to end. More than 350 treaties had been made between the United States and Native American nations between 1776 and 1871, the year that Congress and President Ulysses S. Grant abolished all future treaty making between the federal government and Indian nations. The New Colonialism had arrived. Future treaty abolition came through a rider to the Indian Appropriation Act of 1871. It was straightforward and direct and provided an important guarantee:

> No Indian nation or tribe within the territory of the United States shall be acknowledged or recognized as an independent nation, tribe or power with whom the United States may contract by treaty: Provided further, That nothing herein contained shall be construed to invalidate or impair the obligation of any treaty heretofore lawfully made and ratified with any such Indian nation or tribe.[28]

Why did the United States government take this rather unusual step—that of abolishing future treaty-making powers? A number of explanations, including some with rather narrow reasoning, have been postulated by scholars. The Resolution of 1871 has been interpreted as the result of a desire by the House of Representatives to infringe on the Senate's traditional foreign affairs powers; as a congressional melding of fiscally conservative sentiment, anti-Indian attitudes, and reformer conversion to the idea that the treaty system unfairly took advantage of Indian parties; or as Congress's blind willingness to follow the recommendations of the Indian Peace Commission of 1867/1868 and the Board of Indian Commissioners.[29]

These reasons are certainly plausible for the immediacy of the treaty abolition decision, but the status of Indian diplomacy with the United States and with other tribes during the 1860s also is crucial to any understanding of this massive rupture in diplomatic relations. During the 1860s, the United States accelerated its treaty-making overtures to Indian nations. Countless councils were held, and fifty-nine treaties were actually ratified; indeed, the 1860s represented the most intense era of Native American–United States treaty making ever. The Great Plains peoples were extensively involved during this decade, alone accounting for over one-third of all treaties made in the 1860s.[30]

The largest number of treaties, however, were made with Sioux peoples. During the 1860s, eleven treaties with the different divisions of the Sioux were signed, the most significant being the Treaty of Fort Laramie of 1868, between the United States and the Brulé, Yanktonai, Oglala, Miniconjou, Hunkpapa, Blackfeet, Cuthead, Two Kettle, Sans Arcs, and Santee Sioux. In brief, the Fort Laramie proposals provided for a permanent peace after twenty years of sporadic United States–Sioux warfare. It stipulated that all whites who had committed wrongs or depredations on Indians would be punished, and it established the Great Sioux Reservation, an area covering the entire portion of South Dakota west of the Missouri River. Clauses in this treaty also dealt with the establishment and operation of the reservation. Any Indian who was head of a family was permitted 320 acres for farming. The treaty further stipulated that the country north of the North Platte River and east of the Big Horn Mountains was unceded Indian territory. No one could settle there without permission of the Sioux. The United States agreed that within ninety days after the treaty was signed, all military posts currently established in Sioux territory would be abandoned and that the Indians in return would withdraw all opposition to railroads and travel roads south of the North Platte. Spotted Tail and many other chiefs signed the treaty on April 29, 1868; one important leader, Red Cloud, refused, but finally on November 6, 1868, he signed and the ratification process began.[31]

Both the Treaty of Fort Laramie and Red Cloud's hesitation were instrumental in the decision of the United States to change the current treaty-making process. The government and military had already begun to have serious doubts about dealing with Red Cloud and the Sioux. Contrary to the usual United States policy of opposing Indian resistance, they conceded to Red Cloud everything for which he had been fighting. It was a retreat for the United States and a victory for the Sioux.

In 1870, Red Cloud traveled to Washington, D.C., and it became evident during his conversations with government officials that his understanding of the Treaty of Fort Laramie was completely unrelated to the actual document. When this was brought to his attention, Red Cloud observed,

> When you send goods to me they are stolen all along the road, so when they reach me there is only a handful. They held out a paper for me to sign and that is all I got for my land. I know the people you send out there are liars.[32]

He was despondent to learn that the Sioux were not to be supplied with ammunition and horses, and the Sioux believed that Red Cloud had been deceived. In Sioux country, it became increasingly clear to all that the old treaties would not satisfy miners' and settlers' demands for access to the Great Sioux Reservation. Rather, reductions in Sioux homelands would follow only after a bloody confrontation.

Red Cloud and the Sioux nation had a direct impact on the prevailing colonial policy, as did the other Plains Indian nations that rejected American overtures to them. These troubles caused the federal government to lose faith in its own abilities to practice Indian diplomacy. Moreover, many in Congress became convinced that the government's involvement with treaties simply was not worth the price paid. This malaise was a key element in the abolition of the treaty-making process.

Destroying the Land Base and Forced Acculturation

Despite the Resolution of 1871, past treaties seemed indestructible. That is, the degree of sovereignty held by Indian nations at the time of an agreement could not distort the legal equality inherent in the treaty. Even the American courts recognized this notion. "It is contended that a treaty with Indian tribes, has not the same dignity or effect as a treaty with a foreign and independent nation," a federal circuit court in *Turner v. American Baptist Missionary Union* (1852) summarized. "The distinction is not authorized by the Constitution."[33] Consequently, other legal means had to be employed in order for Indian homelands to be dismembered, both literally in the form of lands taken out of the hands of Native Americans and figuratively with the lessening of the legal power residing in the treaty.

At the time that the House of Representatives adopted the Resolution of 1871, Congressman Henry L. Dawes of Massachusetts argued against abandoning the treaty system. Dawes was especially angered by remarks made by Congressman Thomas Fitch of Nevada, who pressed for the abolition of future treaty making because he believed that the treaty system was wasteful and corrupt, which clearly it was. But Fitch also postulated that the natural extermination of Native Americans would be best for all. To these remarks, Dawes retorted, "As between butchery and humanity, I prefer humanity."[34]

In 1871, Congressman Dawes believed past treaties between the United States and native peoples to be an imperfect yet necessary system, but by 1887 Dawes, now in the Senate, had changed his mind and considered the treaty system to be no longer viable. To Dawes, the Indians had to be changed or "reformed," and the Indian haters in Congress found common ground with him and other "reformers." This political realignment spawned the Dawes Severalty Act, or General Allotment Act of 1887, which codified what had been taking place on a piecemeal basis for two decades.[35]

Essentially, the General Allotment Act broke up the landmass of Native Americans. Under this law, the president could authorize setting aside portions of reservations to individual Indians. Heads of families received 160 acres, and other individuals were given 80 acres. If the lands to be divided were suitable only for grazing, the size of the allotments would be doubled. Later, Congress cut these proportions in half.

Of particular importance were three sections of the act. One allowed the sale to non-Indians of what the federal government termed "excess lands" from the reservations after the allotments had been assigned. Another provided title in trust to the allotments to be held by the United States for twenty-five years or more. This was designed to help protect individual Indian allottees from losing their land. However, the act provided for state or territorial laws of descent and partition to apply, so that Indian families that included Indians married to non-Indians might lose their lands because of divorce or death. The third section made those Indians who accepted allotments citizens of the United States. They could also become citizens if they agreed to separate themselves from their cultural past.

The act included, as well, sections allowing Indian agents or specially appointed agents to allot the lands. Individual Native Americans were supposed to be able to make their own choices, if practicable. Other sections dealt with the water rights of allotments, Indians living off reservations being able to choose an allotment from reservation lands, and the benefits that Indians were to receive. Specifically exempted from the act were select tribes, such as the Cherokees and Choctaws in Oklahoma, the Seneca Nation in New York, and one reservation in Nebraska created not by treaty but by executive order.[36]

The implementation of the General Allotment Act for Native Americans was a disaster. Through leasing, many Indians lost the use of their allotments. Real-estate agents duped Indians into long-term leases, and the non-Indian occupiers then stripped the land of its natural resources, particularly timber. Guardians appointed for Indian minors grossly abused their trust, some even to the extent of murdering their wards. Indians who made out wills leaving their allotments to non-Indians suddenly and mysteriously died. Child marriages were arranged between non-Indians and minors.

In addition to the human suffering, most of the tribal lands were lost. Between 1887, when the General Allotment Act was signed, and 1934, when reforms under the Indian New Deal put an end to allotment, Indian lands decreased from 138 million acres to 48 million acres. Of the remaining 48 million acres, nearly half were desert lands. The process itself accelerated once Indians obtained general citizenship in 1924.[37]

While their land base was being sufficiently weakened, Indian nations also felt the wrath of the federal government, which ruthlessly pursued the forced acculturation of native peoples. The process by which one nation's culture is changed by exposure to another culture and by which cultural borrowing and sharing take place without coercion is called *acculturation*. During the last third of the nineteenth century, the United States used physical force and legal intimidation to induce Indians to reject their own cultural values in a process called *forced acculturation*.

Law was employed by the federal government in its drive toward forced acculturation. Many non-Indians believed that once the new policies were sufficiently established, cultural change would accelerate, and Indians would cease being Indians. Individual Indians soon found they could not obtain the protection of the Bill of Rights. This crucial aspect of the New Colonialism began with efforts by Congress to concentrate all powers over Native Americans in the office of the Commissioner of Indian Affairs and his agents. This increase in grants of power to the BIA, or Indian Service, was accompanied by a corresponding diminution of Indian rights.[38]

The same Appropriation Act of 1871 that forbid future treaties severely restricted the rights of most Indians to make contracts. All contractual agreements in the future had to be approved by the Commissioner of Indian Affairs and the Secretary of the Interior. The commissioner exerted economic control over all tribes in his jurisdiction. Abuses occurred almost immediately, and Native Americans quickly found that this provision added a further restriction: They were deprived of the right to a free choice of counsel. Indians could not retain an attorney to sue an agent or the Commissioner of Indian Affairs without the permission of the very persons named in the suit. Similarly, if an Indian was a defendant in a criminal lawsuit, his or her choice of counsel had to be cleared with Indian Service officials. The Sixth Amendment was, in effect, put on hold for Native Americans.[39]

The linchpin of this movement to consolidate power over Indians in order to carry out the New Colonialism was a closed accounting system. Every Indian agent was authorized to distribute supplies only if particular Indians were listed on rolls maintained at the agencies. These rolls included the names of Indians, heads of families or lodges, and the number in each family or lodge. Furthermore, supplies were not to be given out more than one week in advance, nor were they to be distributed by a headman or leader of a tribe or band. Only the agents were vested with the power to control all food distribution.[40] Once the Indian Service controlled the basic elements of survival, various forms of acculturation could be pursued.

The 1880s ushered in an era of profound change for most Native Americans. This was a time in which the full brunt of the Unites States legal system struck. The federal government pursued forced acculturation policies with abandon. Either Bill of Rights considerations were simply not a matter deemed important, or they were brushed aside.

The right to assemble, protected in the First Amendment, was of particular concern, especially to the Commissioner of Indian Affairs. As early as 1872, Commissioner Francis A. Walker viewed the reservation as a prison in which Native Americans were to be forced into a non-Indian culture. "Indians should be made as comfortable on, and as uncomfortable off, their reservations as it was in the power of the Government to make them," he wrote. He then urged his agents to harass and "scourge" any Native Americans off their assigned space. "Such a use of the strong arm of the Government," according to Walker, "is not war, but discipline."[41] Some Indians were eventually allowed to leave the reservation, but they had to obtain licenses, which required permission all the way to the commissioner's office.[42] A policy statement from the 1890 census summarized the erosion of this particular First Amendment right: "The Indian not being considered a citizen of the United States, but a ward of the nation, he cannot even leave the reservation without permission."[43]

By the 1890s, the courts were beginning to look at some of these restrictions on Indian travel. One federal court noted that reservation Indians were treated "little better than prisoners of war."[44] Earlier, in *Tully v. United States* (1896), the same court had cried out at the treatment of Apaches in Arizona, where any who left their desert reservation were hunted down and killed "as if they were wild animals."[45] But these pronouncements did little to erase the impact of the Massacre at Wounded Knee in 1890 when Big Foot's band of Miniconjou Sioux were murdered by the United States Army. It was an important reminder to any Native American who doubted the efficacy of trying to exercise his or her right to assemble off the reservation.

The 1880s also brought significant legal developments in the area of Native American criminal rights and possible Fifth Amendment ramifications. Indian agents wished to gain further control over the reservation, and the means toward this end was through police powers and local courts. In 1878, the first Indian police force, the Light Horse of the Oklahoma Creeks, was created through congressional funding, and by 1884 two-thirds of all Indian agencies had an Indian police force. The Indian police became the enforcers of forced acculturation policies. They rounded up children to send away to off-reservation boarding schools. They performed public-works jobs for some agents, and they took the census. They wore non-Indian clothing, cut their hair, and took individual allotments of land once they were authorized. When the Indian police were sent to arrest Sitting Bull, they killed him. Thus the Indian police, although staffed by Native Americans, were regarded as bitter enemies by most reservation Indians.[46]

In 1883, the Indian police were made into judges. This curious melding of executive and judicial functions came about with the creation of the Courts of Indian Offenses. Judges were to arrest as well as convict Native Americans who practiced polygamy, who attempted to perform the traditional duties of a medicine man, and who performed various dances or other ceremonial functions of Indian religion. Police judges were, in essence, authorized to put into place the administrative determinations of the Indian agents, the Commissioner of Indian Affairs, and the Secretary of the Interior as they acted as justices of the peace and probate courts.[47]

In the same year that the Courts of Indian Offenses were being established on reservations, the Supreme Court issued a significant ruling in *Ex parte Crow Dog* (1883).[48] Crow Dog had murdered Spotted Tail, a noted leader of the Brulé Sioux and head chief of the entire Sioux nation. Although Crow Dog, according to Sioux custom, paid the relatives of Spotted Tail $50, eight horses, and one blanket as restitution for having committed this act, the United States District Attorney for Dakota Territory believed that Crow Dog should be prosecuted in federal court. Crow Dog therefore was arrested by the Rosebud Indian agent, and a federal district court jury convicted Crow Dog of murder in the first degree. The district court judge sentenced Crow Dog to hang, and the Dakota Territory Supreme Court upheld the conviction.[49]

On appeal, the United States Supreme Court reversed the Dakota Territory courts. In his majority opinion, Justice Stanley Matthews found that federal district courts had no jurisdiction over Indians for murders or other crimes committed by Indians against other Indians on Indian reservations. Matthews noted that through treaties, the United States government had recognized Indian tribes as semi-independent nations. Members were exempt from federal laws because of prevailing customs recognized in the treaties, and no federal laws existed to counter these customs. Only Congress could change this situation, by providing new laws of criminal procedure.[50] Congress quickly responded with the Major Crimes Act of 1885, which made it a federal offense for an Indian to commit murder, manslaughter, rape, assault with intent to kill, arson, burglary, or larceny against another Indian on an Indian reservation.[51]

The Major Crimes Act and subsequent acts have not denied Indian courts jurisdiction over other criminal acts related to the commission of a major crime, and this has caused Fifth Amendment problems. For example, if an Indian committed a petty assault on an Indian that he raped, first the Court of Indian Offenses could try the perpetrator on the petty assault complaint, and then a federal court could try the same Indian for rape. When this did happen later in the twentieth century, the Supreme Court ruled that the double-jeopardy provision of the Fifth Amendment did not apply to prevent the two trials.[52]

The enforcement of the Major Crimes Act, the creation of the Courts of Indian Offenses, and the establishment of the Indian police were important developments in the acculturation movement. Indians on reservations were under significant scrutiny, and local police powers and judiciaries sought to impose a new culture on them. The pressure generated was noteworthy because these new institutions not only cut into the marrow of the essential cultural practices of native peoples, but also constituted a direct attack on fundamental rights. These agencies of procedure easily became active agents of the New Colonialism, and their efforts were confirmed by the Supreme Court in *Talton v. Mayes* (1896).[53]

In this case, a man named Talton murdered another Cherokee in Cherokee country. He was tried and convicted by a Cherokee court after a five-member Cherokee grand jury indicted him for murder. Talton appealed to the Supreme Court, arguing that his Fifth Amendment rights had been violated. He asserted that no grand jury by Fifth Amendment definitions had met (a federal grand jury consisted of a minimum of six jurors) and therefore that he had been wrongfully convicted of this crime. The Supreme Court did not accept this argument, with Justice Edward White writing in the majority opinion that Indian courts were not federal courts.

Justice White admitted that all Native American courts were subject to federal regulation, but he insisted that regulation alone did not make them federal courts subject to Fifth Amendment limitations. To White, tribal powers were inherent, derived before the existence of the Constitution. That the federal government could create new courts and new institutions on reservations did not make those legal institutions exclusively federal. Thus protection against illegal indictments, double jeopardy, and self-incrimination were denied to Native Americans. So, too, were guarantees of due process and the taking of property without just compensation.

In this atmosphere, total administrative control was vested in the Indian agent. Before the Civil War, Indian agents were dominated by frontier politicos who dabbled in small-scale and sometimes large-scale corruption. But after 1871, the Indian agent became the focal point for the enforcement of acculturation rules, and much interest in the office ensued. During the administration of Ulysses S. Grant, a new policy, the so-called Quaker policy, was implemented. The agents chosen were persons who had strong religious backgrounds and who were committed to the destruction of Native American culture through Christianity.[54]

By turning the Indian Service over to Christian zealots, the federal government sanctioned the mixing of church and state on most reservations. All the powers of the Indian agent were channeled toward using Christianity to destroy Indian culture, especially Indian religious practices. Both the establishment and the free-exercise clauses of the First Amendment were blatantly compromised.

The Christian religion was established with federal funds on Indian reservations through education. Mission schools were constructed, and Indian children were required to attend them. Rivalries developed among various Protestant denominations and between Protestants and Roman Catholics over school funding. This competition moved into the political arena, where Congress and the executive tried to remedy the problem by giving virtual denominational monopolies on specific reservations. For example, Catholics were given control of education and the agency at the Rosebud Sioux Reservation in South Dakota, although Catholic-school funding at Rosebud was challenged by Reuben Quick Bear, a Sioux Protestant. The Supreme Court, however, ignored Quick Bear's argument that the educational system amounted to an establishment of religion prohibited by the First Amendment. Instead, the Court used another portion of the amendment to justify its actions. In an especially strange form of reasoning, the majority opinion in *Quick Bear v. Leupp* (1908) held that to cut off these funds to the Rosebud Sioux Reservation constituted an infringement on its First Amendment rights to the free exercise of religion and the ability of Catholics to oversee this religious option. The Indians' First Amendment considerations were dismissed.[55]

The Court's true feelings about Christianity as an active agent of the New Colonialism were revealed in 1894 when another dispute brought Indians to the Supreme Court. The Osage protested Congress's granting railroad right-of-ways to the Missouri, Kansas, and Texas Railway Company across the Osage Reservation. The Court found for the railroad company against the Osage, and it reiterated a policy that allowed Congress to control Indian lands within the political sphere. But the justices went further, stating that in cases involving Native Americans, "the United States will be governed by such considerations of justice as will control a Christian people in their treatment of an ignorant and dependent race."[56]

By the turn of the century, the legal circle of acculturation policies was nearly closed. Complete power over Native Americans had been placed in the hands of non-Indians. Individual Indians had virtually no recourse to the Bill of Rights to protect themselves. They were trapped on islands of mostly worthless land, and they could not leave. They had lost control of their police powers and local courts, and their Fifth Amendment rights could not be promulgated.

Constitutional precedents could not prevent Indian agents—with all the powers of the law behind them—from attacking, without interference, Indian religion, family life, education, and other basic cultural values. A means of separating Native Americans from their reservation lands had been discovered with the General Allotment Act of 1887. All that remained for total forced acculturation was to attempt to destroy the treaty in which could be found the dormant seeds of legal protection.

Lone Wolf v. Hitchcock *(1903)*

Many tribes resisted allotment and the cultural attacks on their lifeways. They tried to prevent the loss of their land and, with it, their distinctive worldview. The Kiowas were one such Native American nation that sought to protect their reservation in southwestern Oklahoma, a homeland guaranteed by the Treaty of Medicine Lodge Creek, signed by the United States and the Kiowas, Comanches, and Apaches in 1867.[57]

In 1889, Congress authorized the creation of a three-member committee, the Jerome Commission, named after its chairman, David Jerome, to go to Oklahoma to negotiate with the Kiowas and other Indians, who were to select individual parcels of land and sign an agreement regarding the sale of "surplus" reservation lands.[58] The Treaty of Medicine Lodge Creek carried a stipulation that

> no treaty for the cession of any portion or part of the reservation herein described [the Kiowa–Comanche–Apache Reservation in Oklahoma], which may be held in common, shall be of any validity or force as against the said Indians, unless executed and signed by at least three fourths of all the adult Indian males occupying the same.[59]

Thus permission had to be obtained from the Kiowas, Comanches, and Apaches in order to allot reservation lands.

Hearings were held at Fort Sill in September 1892. Many Kiowas came forward to let the Jerome Commission know of their opposition to allotment and especially to the sale of surplus reservation land. An agreement was signed by several leaders, including Lone Wolf, but what they thought it said and what it actually said were quite different. The commission then set about to find three-quarters of the adult Kiowas to agree to give up the surplus lands. By the time the commission tried to do this, the Kiowas knew that the agreement they had originally signed was fraudulent, and they refused to sign the petitions. With the collusion of the Kiowa agent, non-Indians were added to tribal rolls; Kiowas were taken off the rolls; signatures were mysteriously added; and many Kiowas who actually did sign wanted their names stricken.[60]

Although it took nearly seven years, Congress accepted the agreement in June 1900. At this point, Lone Wolf and several other Indians sued the Secretary of the Interior, Ethan A. Hitchcock, to prevent him from allotting the reservation. The Kiowas lost in federal district court in Washington, D.C. The district court judge ruled that it made no difference whether irregularities were involved or if misunderstandings had taken place; Congress had the final say on all matters pertaining to Native Americans. On appeal, the District of Columbia Court of Appeals again ruled against Lone Wolf. This time, the court stated that treaties were not a proper instrument for the court to evaluate and that lands held by Indians were subject to the control of the United States.

This decision was appealed to the Supreme Court. Lone Wolf's lawyers argued that the allotment agreement passed by Congress dividing up the Kiowa–Comanche Reservation was unconstitutional because it violated the Treaty of Medicine Lodge Creek, because it was a fraudulently conceived document, and because the Kiowas' property rights had been deprived without due process of law under the Fifth Amendment.

Justice Edward White wrote a near-unanimous opinion. He upheld the lower courts and went further. To White, Congress had unlimited power over Indian property, regardless of any treaty guarantees. Why might Indian treaties not be valid legal instruments? White concluded that there were times when emergencies required the treaties to be ignored, and the absence of Indian consent could also be ignored. He reasoned, too, that the Resolution of 1871 could be interpreted to mean that Congress not only was prohibited from making future treaties, but also had the power to abrogate any past treaties. White refused to consider any Fifth Amendment implications.[61]

Lone Wolf v. Hitchcock was devastating, as it justified the unilateral termination of treaties. No longer might Indian nations attempt to protect themselves. No longer could any semblance of rights be claimed under the Constitution based on treaty guarantees. The key to the alienation of Indian homelands—the abrogation of treaties—had been found. The New Colonialism was secure. It had been safely ensconced in the executive through Indian agents; it had been given a statutory base by the legislature; and, finally, it had been provided with a suspect constitutional justification for its programs by the judiciary. To many interested non-Indians, forced acculturation was king, and the Bill of Rights had been declared inoperative for Native Americans.

■ ■ ■

In *The Way to Rainy Mountain,* N. Scott Momaday tells the story of an old Kiowa man, his wife and child, and their meeting with an enemy Indian. During the evening, the woman was preparing meat, and she gave a piece to her son. He went outside to eat it, and when he returned he asked for more. This happened three times, but when the boy came in again he was accompanied by an enemy who told the family that they were surrounded by enemies who only wanted food. He explained, "If you will feed us all, we will not harm you."[62]

The old man was suspicious, but he pretended to cooperate. His wife put fat on the fire while he went outside to hide his enemies' horses. Momaday wrote,

> When he was well away, he called out in the voice of a bird. Then the woman knew it was time to go. She set fire to the fat and threw it all around upon the enemies who were sitting there [in the tipi]; then she took up the little boy in her arms and ran upstream.[63]

The Kiowa family escaped, but they could see the fire and hear the screams of the intruders.

During the last third of the nineteenth century, Native Americans surely must have felt that they were surrounded by an enemy, an enemy who sought their homelands, their economy, their children, and their heritage. Beginning with the Trade and Intercourse Acts and culminating with the Resolution of 1871 and the *Lone Wolf* decision, United States law was shaped in order to impose first an Old and then a New Colonialism on Indians. This was a formative era in the legal history of Native Americans, for it provided benchmarks on which the legal debates and actions of the twentieth century could be based. The assault on treaties and Native American rights had begun, but it did not prevail.

3

Legal Survival, Human Survival

The nineteenth century was a time of cautious introductions for Native Americans to the American legal system. There was minimal sparring over rights within the federal Bill of Rights framework. That should not be surprising, as there were few situations when the Bill of Rights was interpreted by the highest court in the land in the nineteenth century for any Americans. Still, Indians rarely were in court, and even more infrequently were they able to pursue grievances to the Supreme Court level. When they did—such as in *Cherokee Nation v. State of Georgia* (1831), *Worcester v. Georgia* (1832), and *Lone Wolf v. Hitchcock* (1903)—the results were extremely destructive.

Indeed, the *Lone Wolf* precedent marked a watershed in Indian legal rights. The doctrine of plenary power over Indian affairs vested in Congress foreclosed the federal Bill of Rights as an available avenue for legal redress. Native Americans were at the mercy of the political system without the traditional checks of the judiciary.

The immediate post–*Lone Wolf* era, the first three decades of the twentieth century, proved to be a time of distress and disaster for Native Americans. Through the guise of reform, the federal government sought to strangle the life breath of Indian culture. No aspect of Native American life was safe. Laws and their execution were renewed and fine-tuned in extraordinary efforts to restrict Indian rights. Of course, the New Colonialism of the previous decades, only recently discovered, offered precedents for the new era. For Indians to survive as native peoples hung in the balance.

This aggressive approach to acculturation is perhaps best described through a comparison of two United States Supreme Court decisions. In 1876, the Court held in *United States v. Joseph* that the Trade and Intercourse Act of 1834 did not apply to the Pueblo Indians of New Mexico because the Pueblos came under the jurisdiction of the United States as a result of the Mexican War, and the United States had not made treaty agreements covering this matter with the Pueblos. Moreover, the Court found that the Pueblos were a "peaceable, industrious, intelligent, honest, and virtuous people. They are Indians only in feature, complexion, and a few of their habits."[1] The Supreme Court accepted an assessment of the Pueblos made in 1869 by the New Mexico Territory Supreme Court, which determined that these New Mexico Indians were "the most law-abiding, sober, and industrious people of New Mexico."[2] Thus since the Pueblos were not acting like other Native Americans, reasoned the Court, they were not subject to laws restricting Indians.

Although the vote of confidence may have been appreciated by the Pueblos in the nineteenth century, all that remained was for any future Supreme Court to change its mind about their character. It did so in 1916. A situation arose in which a person named Sandoval sold liquor to Pueblos on their lands. This was contrary to federal law, which prohibited such sales. A lower court ruled that Sandoval was allowed to do this because the homelands of the Pueblos were not considered Indian country. It cited *United States v. Joseph* as a controlling precedent. The case was appealed to the Supreme Court, which reversed the lower court. In *United States v. Sandoval* (1913), the Court in effect overruled *United States v. Joseph* by determining that the Pueblos' character had changed and that they were Indians after all. Wrote Justice Willis Van Devanter, a westerner known for his strong racist views:

> The people of the pueblos, although sedentary rather than nomadic in their inclinations, and disposed to peace and industry, are nevertheless Indians in race, customs, and domestic government. Always living in separate and isolated communities, adhering to primitive modes of life, largely influenced by superstition and [fetishism], and chiefly governed according to the crude customs inherited from their ancestors, they are essentially a simple, uninformed and inferior people.[3]

The highest court of the United States decided that Indians had certain traits that the law recognized as undesirable and through which they could be destroyed. Native American property rights, religion, family traditions, political institutions, and cultural practices were now defined as "crude customs" worthy of extinction at any price over any legal objections, including those rights guaranteed in the Constitution and the Bill of Rights.

Distress and Disaster:
Enduring Frontal Attacks on Native American
Culture, 1904–1928

By the 1930s, the legal rights of Native Americans had plummeted to new depths. Essentially, no Indian rights were left. During the first two decades of the twentieth century, the federal government ensured this end result. Bureaucratic controls mounted throughout this era. No aspect of Indian life was immune from intervention. State governments became bolder in their approach to Indians, and they were encouraged by a federal bureaucracy to aim regulatory schemes directly at Native Americans.

The statutory base for this approach was the General Allotment Act, or Dawes Severalty Act, of 1887. Inherent in this revolutionary document was the supposition that Indians needed changing. As guardians of the wards, the federal government rapidly expanded BIA operations. Some non-Indians observed that there were problems with this approach, but they were eventually assuaged by the idea that citizenship for all Indians would solve everything. During this era, national citizenship became a reality for Native Americans, but it did not prevent non-Indians from taking Indian lands, from assuming Indian mineral and water rights to those lands, or from continuing the suppression of Indian culture.

The Heyday of the Federal Bureaucracy

The first two decades of the twentieth century were marked by a concentration of administrative power over Native Americans in the Department of the Interior. The Secretary of the Interior and the Commissioner of Indian Affairs were the recipients of administrative blank checks. Especially subject to regulation and attention were the powers to dispose of native lands. The Dawes Severalty Act ensured the Interior Department of this focus.

From 1900 to 1909, Congress refined the administration of the Dawes Severalty Act. Individual leasing of allotments was allowed for the first time in 1900.[4] Then in 1901, the Secretary of the Interior was authorized by Congress to allow easements over allotments and tribal lands without Native American consent. States and territories were also given the power to condemn allotted lands under state eminent-domain legislation. Federal officials did not consider the Fifth Amendment to apply to Indians. Large numbers of acres were thus taken, without proper compensation, from tribes and individual Indians for such public needs as telegraph and telephone lines and roads.[5]

Condemnation procedures were lengthy matters, and they were restricted to public works of a sort. Those desirous of Native American lands wanted easier access, and so they worked toward that end. In 1902, Congress allowed the heirs of a deceased allottee to sell the allotment if the Secretary of the Interior approved of the sale.[6] This provision went against the express purpose of the Dawes Severalty Act by making it possible to take lands out of the hands of individual Indians. Within a few years, thousands of acres were being exchanged, and allottees were disappearing or being murdered in order to expedite conveyances. This, as historian Angie Debo discovered, set off a bloody frenzy of corruption, violence, and land speculation in Oklahoma.[7]

By 1906, Congress had become somewhat distressed over the loss of lands by allottees, but it did not look at the issue very carefully, and after it passed new legislation, the situation was only exacerbated. Congress's primary contribution to the confusion was to pass a bill introduced by Congressman Charles Henry Burke of South Dakota. Under the Burke Act, the granting of citizenship to all Native Americans was delayed, and their allotments were declared to be inalienable until the trust period was over.[8] The length of the trust period could be determined by the president. At first glance, this appeared to halt the sale of allotments. By declaring the lands to be inalienable—in other words, not subject to sale—and by making sure that citizenship was postponed and could not be used as a means of asserting power to sell land, Congress sent a signal to the Secretary of the Interior to provide more scrutiny over land-distribution efforts.

For Native Americans, citizenship was a complicated issue, as they were viewed by American courts as not being citizens of the United

States or being capable of attaining citizenship through normal procedures. After the Fourteenth and Fifteenth Amendments were ratified, some legalists thought that Native Americans were now full-fledged citizens, but subsequent court interpretations held that the Fourteenth Amendment citizenship provisions were not applicable to Indians.

All of this reached an extreme point when in *Elk v. Wilkins* (1884), the Supreme Court concluded that John Elk, an Omaha Indian farmer who had abandoned his tribal membership and farmed his own land, could not assume the position of a United States citizen and exercise voting privileges. An Omaha, Nebraska, suffrage registrar, Charles Wilkins, refused to register Elk because Elk, according to Wilkins, was an Indian and therefore not an American citizen.[9] Native Americans were legally defined as noncitizen nationals with an allegiance to their tribe, unless Congress ruled differently. This ruling could be overcome only by treaty or special statute.

If a Native American somehow managed to become a United States citizen, he or she would receive a certificate from the Secretary of the Interior that empowered him or her to be able to sell real property and to be taxed for that property. By postponing any citizenship accruing after the allotment of lands by the Dawes Severalty Act, such allotments, some reformers believed, were prevented from falling into the hands of speculators. Indians could not control them, and taxing authorities could not get their hands on allotments after failing to pay real-estate taxes on them.

But even with these supposedly delaying provisions, Congress modified its intent. In the same legislative session in which the Burke Act was passed, Congress determined that allotment lands could not be used to satisfy individuals' debts, nor could money earned from leases—unless expressly authorized by the Secretary of the Interior—be applied to such debts.[10] This legislation had important side effects. Indians who incurred debts could not use the only asset they had to overcome their liabilities, but even more important, the lands remained clear of claims, which made the allotments even more desirable to the land-hungry speculators than before.

It did not take long to make these protected lands available to non-Indians. The very next year, the Secretary of the Interior was given new powers to use his discretion and flexibility to design new regulations that would allow allotment sales before the end of the trust period. The Commissioner of Indian Affairs was to supervise all sales.[11] If the looting of native lands was not sufficient, the Secretary of the Interior was also directed to identify individual Indians who were capable of managing their own financial affairs. These fortunate Native Americans were given shares of tribal funds held in trust, and guardians of Indians who were ascertained to be mentally or physically impaired were also given the power to draw on tribal resources.[12]

Congress thought that it was handing over to the Department of the Interior the means to get the government out of the "Indian business." The Interior Department assisted in the separation of Native Americans from their allotments and guaranteed that those lands were not subject to liens or other claims. All those holding claims to Indian indebtedness could deplete tribal trust funds. Any pretense left over from the Burke Act to hold lands and tribal resources for Native Americans was gone.

The final blow to the Burke Act came in 1908. Congress repealed its restrictions on inalienability when it authorized the Secretary of the Interior to allow the sale of allotments after the death of an allottee. The secretary could also give a patent to any heir or any person to whom the heirs had sold the rights to the allotment.[13] An anticlimax came the next year when the secretary was allowed to lease Indian lands to railroads, mining corporations, and irrigation companies for long-term agreements that were "for the best interests of the Indians." Although Native American opinion on the matter needed to be consulted, it could be ignored.[14]

The loopholes available to Indians, few that they were, closed after congressional legislation passed in 1910. The Secretary of the Interior was given the power to manage and sell allottees' estates, including timber. The secretary also was authorized to approve all wills made by Indians with allotments. He could set aside lands for children ignored in the wills and then sell surplus lands to non-Indians. And he could lease individual allotments for periods not to exceed five years and spend the money earned from these leases in ways that he saw fit.[15]

Thus administrative control over Indian allotments was clearly established in the federal bureaucracy. The *Lone Wolf* doctrine gave Congress the ability to place extraordinary powers in the hands of the Secretary of the Interior and the Commissioner of Indian Affairs. By so doing, the Fifth Amendment to the Constitution was bypassed. Native Americans were deprived of property without due process of law, and their private property was taken for public use without just compensation.

The Ultimate Solution: Citizenship

The first decade of the twentieth century marked a time of federal attack on Native American rights in the name of reform. It was relentless. Few areas of Indian life remained untouched. The subsequent decade continued this trend, and the non-Indian leaders of this movement thought that they had solved the "Indian problem" once and for all by the time they passed the Citizenship Act of 1924.

Congress bestowed on the Department of the Interior further powers over Native Americans' lives. In 1913, Congress extended the Secretary of the Interior's oversight of Indian wills to include all Indian testamentary dispositions of rights to Indian tribal funds. Previously, the secretary's approval had been required only for Native American wills that concerned allotments. Virtually all Indians who were members of tribes had to receive the secretary's approval for the validation of their wills.[16]

The Indian Department's Appropriation Act of 1914 prevented all Indians from signing contracts that might affect tribal funds or property without the permission of the Secretary of the Interior. The same act allowed the secretary to quarantine reservations if their residents had any contagious diseases.[17] A similar law passed two years later enabled the secretary to lease Indian allotments to non-Indians for irrigation purposes if the Indian owner could not physically occupy the land. Supposedly this statute applied only to those Indians who owned the land but were underage or disabled, but it proved to be a particularly onerous regulation fraught with great potential for abuse and corruption.[18]

During World War I, although Indians could not be drafted, they were encouraged to join the war effort. Several all-Indian units were recruited, principally for scouting purposes, a move that was later expanded during World War II. The reward for Indian veterans who survived World War I was citizenship, authorized by Congress in 1919. When the new citizen Indians returned home, they discovered life on the reservation and allotments to be worse than it had been before the conflict. The war resulted in cost-cutting measures for Native Americans on reservations. Federal funds were cut for the education of any Indian children who had less than one-quarter Indian blood and whose parents were United States citizens. The same law restricted the creation of new reservations.[19]

In line with the war effort, Congress completed the process of giving the Department of the Interior virtually unlimited jurisdiction over Indian resources. The secretary was authorized to isolate all tribal funds held in trust and to distribute them to each tribal member. In addition, he was given the power to lease without restriction Indian tribal lands not allotted. The secretary also was encouraged to provide very favorable leases to mineral companies. Tribal permission or consultation was no longer required.[20]

The door was now completely open for dissolving Indian tribes, dispersing Indian tribal funds, and destroying Indian control over their lands. Bureaucrats controlling the fate of Native Americans were confident they could do the job; political and legal interference was not a factor. A former Commissioner of Indian Affairs perhaps summed it up best in 1919. "The Indian problem," stated Francis E. Leupp, "has now reached a stage where its solution is almost wholly a matter of administration."[21]

The 1920s were a time of fear and reaction in the United States. It was the time of the Red scare, the Palmer raids on allegedly subversive resident aliens, and a revival of the Ku Klux Klan. World War I created more problems than it resolved. People's lives were fundamentally changed by their new access to places beyond their own milieu, via automobiles and radios. Indians were not immune from these influences, although they had far less contact with them, and they, too, were the subject of reaction.

In 1920, the Department of the Interior was instructed by Congress to force Indian children into federal or state schools.[22] The states resisted this measure. When Alice Piper, a fifteen-year-old California Indian, wanted to go to public school at Big Pine, she was denied admission solely because she was an Indian. She applied for a writ of mandamus to force the school district to allow her to receive an education, and the California Supreme Court ordered her admission. The Fourteenth Amendment was applied, along with state constitutional requirements that children must receive an education.[23] Other states, such as Oregon and Montana, that tried to prevent Native American children from attending state public schools also lost in their own highest courts.[24] Nevertheless, many more Indian children were permanently denied admission to state schools than were enrolled, and finally the federal government gave up and increased its dependence on boarding schools for Indian education.

In 1921, the same general legislation as passed in 1916—by which the Secretary of the Interior could lease, without restriction, allotments to non-Indians for irrigation purposes—was extended to allow leases for general farming purposes.[25] In 1924, oil and gas leases were authorized in a public auction to be conducted by the Department of the Interior.[26] And the citizenship odyssey for American Indians was concluded in 1924 when Congress declared all noncitizen Indians born within the territorial limits of the United States to be citizens of the United States.[27] Previously, Indians had received citizenship through a variety of haphazard means: by receipt of an allotment, by separation from one's tribe, by special permission of the Secretary of the Interior, or by service in World War I, to name a few of the more common routes taken.

The push for this blanket reform was ardently advocated and symbolized by Robert Valentine, the Commissioner of Indian Affairs in the administration of William Howard Taft. Valentine, a Harvard-educated Greenwich Village settlement-house worker, who knew little if anything about Native Americans, strongly favored the Burke Act. He believed that Indians were ''hiding behind their trust patents'' in order to avoid paying taxes.[28] Valentine wanted

> to place each Indian upon a piece of land of his own where he [could] by his own efforts support himself and his family or to give him an equivalent opportunity in industry or trade, to lead him to conserve and utilize his property . . . rather than to have it as an unappreciated heritage.[29]

With citizenship came responsibility and respectability, thought Valentine and many others.

Conferring federal citizenship on Native Americans, however, did not change many things. Citizenship status theoretically gave Indians the right to vote. But this right was not protected by force or federal statutes, and it was not fully attained until several decades later. Some states, such as Arizona, Maine, North Dakota, and Minnesota, successfully prevented the new federal citizens from voting.[30] They argued that the Fifteenth Amendment could be overcome because Indians did not pay state taxes; they were still wards of the federal government, which precluded them from voting; or they were residing on lands that were not a part of the state for voting purposes.

It was also quickly determined that federal citizenship did not interfere with the special legal place of Indians. They were still members of tribes, and they were still under a trust relationship with the federal government.[31] State citizenship remained elusive. The Fourteenth

Amendment supposedly guaranteed state citizenship to Native Americans if they were citizens of the United States, but the wardship concept offered alternative interpretations. Thus the culmination of legal maneuvers giving Indians citizenship was greeted with a legal yawn. About all it meant for Native Americans was that they could be taxed and lose still more property.

Water Rights and the Winters *Doctrine*

In the midst of the creation of regulatory powers over Native Americans and their lands came a curious legal dispute. The issue litigated concerned the extent to which Indians had interests in water rights. Certainly Indians needed water. They were living on some of the most remote and parched real estate in the United States, and the basis for much of the regulatory power of the Department of the Interior and the concept of wardship required that Indians have some means to become non-Indians. Thus where Native Americans fit in with others who wanted scarce water supplies was important.

Two competing systems of water law developed in the United States during the last half of the nineteenth century. At first, the common-law *doctrine of riparian rights* was accepted wholesale by the eastern states. Riparian rights give to the owner of land on a creek or lake the right to the reasonable use of the water. Such a right is a part of the land; it cannot be sold. It does not make any difference whether or not the right is used, since it cannot be defeated as long as the land exists. Reasonable use depends on how the water is used, and it can be exercised as long as it does not interfere with the flow of water going to other riparian users. Drought decreases usage proportionately. Such a doctrine is especially conducive to humid areas—that is, mainly those lands found east of the Great Plains.

A separate system of water law is more attractive in much of the arid American West, where the *doctrine of appropriation* evolved. It began, in part, because of the mining industry's need for water and because the mines were not located directly on water sources. Agriculture later developed and required irrigation. Thus water rights are not attached to the land; instead, they belong to the first user who puts the water to a beneficial use. An appropriator can reserve the same amount of water taken for a legitimate purpose without having to worry about the rights of later users as long as the same amount is actually used. If one uses less water, one's rights will be adjusted accordingly. Once a stream is fully appropriated or if there is a drought, the last users will lose out.

Competition for western water rights encouraged aggressive behavior by non-Indian farmers, ranchers, and miners. The states were supportive as well, and their enthusiasm was matched by federal agencies, which situated on reservations as many irrigation projects as possible. Tribal trust funds were drained for this purpose, and tribal lands and allotments were leased at very low prices to accomplish this goal. Oddly, this rather nefarious federal action helped Indians retain their rights to water.

In 1908, the Supreme Court heard arguments for the case of *Winters v. United States*. The dispute involved the Fort Belknap Reservation of the Gros Ventres, Piegans, Bloods, Blackfeet, and River Crows in Montana. Lands adjacent to the reservation had been sold to non-Indians as a result of allotments from an 1888 agreement. The compact stated that the boundary of the reservation was in the middle of the Milk River. In the 1890s, non-Indians settled on the land opposite the reservation, filed for water claims based on the Desert Land Act (which stated that all non-navigable waters were available for appropriation, depending on state laws), and began diverting water for irrigation purposes. The Fort Belknap Indians' only water supply was the Milk River, and they sued to prevent its diversion.[32]

The Supreme Court had decided in *United States v. Winans* (1905) that Indians could exercise some rights off the reservations, including treaty rights to fish in all customary places. At issue in *Winans* was Washington State's attempt to prevent members of the Yakima Nation from setting fish traps in the Columbia River. The state argued that the Yakimas had given up title to the lands surrounding their customary fishing places, and by so doing had lost any special treatment reserved in the treaties. The Court, however, rejected this reasoning, because it found that treaties were not grants of rights *to* Indians, but grants of rights *from* Indians. Native Americans may have given away lands, but if they were reserved in the treaty, they did not give away fishing rights with the lands.[33]

Moreover, the Court hinted at the problems caused for Indians by the forced acculturation. As Justice Joseph McKenna wrote, fishing was essential to the survival of the Yakimas, so much so that it was as important as "the atmosphere they breathed. New conditions [such as poverty, lack of food, failure of federal efforts to train or feed Native Americans, and disease] came into existence, to which those rights [fishing at customary places] had to be accommodated.''[34] What the Court concluded was that if Indians were to be successfully assimilated, they must have some fundamental rights reserved so that they could actually move from being Indian to being non-Indian. One of those rights was food. Another might be water.

The Supreme Court picked up on this theme in deciding *Winters.* Looking at the allotment agreement of the Fort Belknap Indians, the Court held that the reservation lands that had been given up were part of the Indians' original homelands and that these homelands had been ceded because of the desire of these particular Native Americans to ''become a pastoral and civilized people.''[35] Since the cession had been made in order to adopt ''civilized'' ways, the Court held that Indians retained water rights in order to prevent their allotments and retained tribal lands from becoming valueless. Montana settlers argued that the Indians had deliberately given up their lands with their water rights, that they had made no effort to reserve any water rights, and that the federal government knew this and understood it.

But the Court did not care about these arguments. ''We realize,'' said the majority, ''that there is a conflict of implications, but that which makes for the retention of the waters is of greater force than that which makes for their cession.'' The Court then turned the language of appropriation into a right for Native Americans. ''The Indians had command of the lands and the waters—command of all their *beneficial use,* whether kept for hunting, 'and grazing roving herds of stock,' or turned to agriculture and the arts of civilization.''[36]

There is much of import in the *Winters* case. One significant observation is that Native American water rights did not accrue from the Fifth Amendment. Nowhere do we see the Constitution or the Bill of Rights used as a foundation for protecting such valuable property as water in the arid West. Instead, Indian water rights evolve from the campaign for acculturation. Because water was essential to the Indians' survival, it had to be reserved for them for their beneficial use, and such beneficial use was very broadly defined. No longer was it imperative for the federal government to lease Indian lands in order to protect Indian water rights. The lands could be left as they were. In time, Native Americans might develop their lands with their preemptive water rights. Moreover, Indians could prevent others from depleting water by merely asserting their prior claims.

The *Winters* Doctrine retains both appropriation and riparian derivatives. First, the rights reserved are federal rights, and they supersede state and local laws. Second, the establishment of a reservation includes not only a "reservation" of lands, but also a "reservation" of Indian water rights in all waters inside or bordering the reservation. These rights to water are held against all competing users, except those who had prior appropriation rights designated under state laws before the creation of the reservation. Some reservations, such as those in New Mexico, date back to the seventeenth century and have virtually no prior users, but for the recently created Yaqui Reservation outside Tucson, Arizona, Indian water rights are only a few years old. Perhaps most important to Native Americans, Indian rights to water were not lost by not using them; they were retained by the tribes for future needs.

Questions still remained after *Winters*. The most significant had to do with the quantity of water reserved for Indian use, but there is no denying that the *Winters* Doctrine was a bold declaration at an incompatible moment in the legal history of Native Americans. State governments realized its significance and refused to recognize it. Senator William Borah of Idaho noted that "the Government of the United States has no control over the water rights of the state of Idaho." Senator George Sutherland of Utah pretended that *Winters* had not happened. He thought it "one of those unfortunate statements that sometimes courts, and the highest court, lapse into."[37] This is not really what the Court meant. Western reaction to the assertion of Native American rights outlined in the *Winters* case proved to be quite similar to southern reaction in 1954 to *Brown v. Board of Education*. And for the Indians and their water rights, it was one thing to hold an established legal right, but it would be quite another to assert it against a hostile world.

Revelations of the Meriam Report

An example of one such hostile senator was Holm O. Bursom of New Mexico, who in 1922 introduced the so-called Bursom bill, which attacked Pueblo water rights and their land base. The proposed law was defended by Charles Henry Burke, who after his stint in Congress had been tapped by President Warren G. Harding to head the Indian Service.[38]

Charles Burke had utter contempt for Native American culture, and he devoted his life to trying to alter Indian beliefs by education and persuasion or, if necessary, by force. One of his first acts after taking office in 1921 was to restrict Native American religious practices. To Burke, "the native dance still has enough evil tendencies to furnish a retarding influence and at times a troublesome situation which calls for careful consideration and right-minded efforts." Addressed to the superintendents of all Indian agencies, Burke's memo urged the suppression of the sun dance and all other Indian "so-called religious ceremonies." Only those dances deemed wholesome entertainment and not religious practice could be held.[39]

The opposition was led by John Collier, a social worker from New York City who worked with the Pueblos and was convinced that they had a viable worldview that worked for them and should not be disturbed. Collier testified at congressional hearings, and he became enraged with what he termed Burke's attempts to destroy Southwest Indian culture. Eventually a compromise law, the Pueblo Lands Act of 1924, was passed by Congress, but the controversy over Pueblo lands alerted the country to Indian issues.[40]

All sides in the controversy wanted information about the status of Indians, but little comprehensive data were available. In order to obtain usable information, Secretary of the Interior Hubert Work turned to the Institute for Government Research (IGR), a private group that eventually became part of the Brookings Institution. In 1926, the IGR agreed to prepare an impartial report surveying the current condition of Native Americans.[41] Chosen to head the investigation was Lewis Meriam, a graduate of Harvard Law School and a staff member of the IGR. Nine specialists in agriculture, health, family life, law, economics, and urban Indian affairs were appointed to assist him. A Yale-educated Winnebago, Henry Roe Cloud, joined in an advisory capacity, and together Meriam and Roe Cloud wrote *The Problem of Indian Administration,* or what became known as the Meriam Report,[42] a document that had profound repercussions.

The approach taken by the Meriam study group was to evaluate conditions for Indians in the United States. Meriam chose specifically not to compare Indians in 1926 with those of earlier time periods. Instead, he wanted to look at current conditions and evaluate them with reference to the job being done by the federal government. Any comparisons were drawn from evaluations of how the Indian Service matched other federal agencies doing similar things for other American citizens. Seven months of fieldwork plus visits to ninety-five reservations resulted in a pamphlet submitted on February 21, 1928.[43]

What did the investigation find? The Meriam Report documented deplorable conditions for Native Americans. In health care, Indians were found to be without basic health services. Hospitals were rare, and the few that existed had minimal equipment, lacked qualified medical personnel, and failed to meet minimum hospital standards. Indians suffered extremely high rates of infant mortality, twice the average for the general American population. Their death rate from tuberculosis was seven times the national average. Many Indians suffered from trachoma. Sanitary conditions were very bad, and medical care for children was extremely deficient. Most Indians were disease ridden.[44]

The Meriam Report scored education as a dismal failure and Indian boarding schools as "grossly inadequate." From 1900 to 1926, the Indian Service made a top priority of separating Indian children from their parents and placing the children in Indian boarding schools. The 26,451 Indian children in boarding schools in 1900 were compared with the 69,892 Indian children confined to boarding schools in 1926. The Meriam study noted the harsh discipline heaped on Indian children, who were not allowed to speak their language, practice their religion, or wear traditional clothes or hair styles for fear of physical abuse. Most schools relied on Indian child labor to make ends meet. The schools were characterized as overcrowded and staffed with unqualified personnel who offered improper medical care and served an improper diet. Indian children at boarding schools were fed at a cost of 11 cents a day. The illiteracy rate of Indians in one state in 1926 was 67 percent.[45]

No economic or legal structure seemed to be in place for Native Americans. The report found that 2 percent of all Indians were earning in excess of $500 a year in 1926, that 96 percent of all Native Americans were realizing less than $200 a year, and that 66 percent of all Indians made less than $100 a year. Forty-nine percent of all Indians had recently lost their land. The legal system was extremely confused, and in this confusion, unscrupulous persons manipulated the law to take advantage of many Native Americans. Authorities were unsure who heard the cases that involved Indians or non-Indians as victims or defendants, on reservations or off reservations, and when they were heard, justice was often not a result.[46]

Having identified a number of problems, what did the Meriam Report recommend? Basically, it urged a fresh infusion of funds to do the job correctly. It suggested new programs and called for the creation of a new office in the Indian Service that would plan new programs and monitor existing ones. The report identified the federal government and the Indian Service in particular as exhibiting an extremely hostile attitude toward Indian culture and Indian families. Moreover, it found the allotment system to be the major cause of the horrible conditions that prevailed. In addition, for the first time, urban Indians were recognized as an important group, and the report specifically recommended that an Indian claims commission of some sort be set up to resolve past legal disputes, particularly those dealing with Indian rights and treaty issues. This last recommendation was regarded by the Meriam study group as an important first step in the legal resolution of any other problems for Native Americans.[47]

What the authors of the Meriam Report prescribed was not radical surgery; rather, it might be more accurately regarded as a "scientific" exposé advocating "moderate" change within existing bounds. Still, it was unusually honest, observed Vine Deloria, Jr., for a government report.[48] James E. Officer terms the report "the most searching study of Indian administration ever undertaken,"[49] and Charlotte Lloyd Walkup, a new assistant working in the Department of the Interior during the early years of Franklin Roosevelt's administration, remembers that the Meriam Report "laid out, in cold blood and in a very terrifying way, what the problems were that needed to be solved."[50]

The Meriam Report documented a national scandal. The forced acculturation of Native Americans as a government policy had failed. To be Indian meant that one was poor, ignorant, and sick—not a "civilized" citizen of the United States. Reservations and their deplorable conditions were simply a product of government policies and government neglect. Blame could clearly be assessed.[51] Might the report have an impact on government policy?

In 1932, Felix S. Cohen, a New York attorney, along with many other idealists, went to Washington, D.C., as a member of the new Roosevelt administration to stimulate change in American society. Cohen was invited by Nathan Margold, solicitor of the Department of the Interior, to help draft new legislation for Indian tribes. Cohen eventually became an assistant solicitor at Interior to head a new project to survey Indian law, which resulted in his monumental *Handbook of Federal Indian Law*. Cohen saw the Meriam Report as providing the spark to "inaugurate a new era."[52]

"The Meriam Survey," wrote Wilcomb E. Washburn, "was the herald of a movement that has continued to our own day which has sought to emphasize education and training as the key to entrance into white society."[53] Entrance into white society would remain a goal. Non-Indians who were not sympathetic to Indian culture believed this just as strongly as before, but they became convinced that change was needed. Many Indians also came to believe this as well. They wanted to be able to use the legal machinery, including what might be available from the Bill of Rights, to retain two crucial elements for Native American survival: an Indian land base and Indian tribal identity.

During Herbert Hoover's administration, the Meriam Report was taken very seriously, and attempts were made to implement some reforms. Particularly offensive Indian Service personnel were forced to resign; some new funds were committed to improving health care; and Indian boarding schools were deemphasized. But there still were no basic changes. No Indian claims commission was approved; no economic plan that included ways for Indians to market their arts and crafts was designed; no tribes were incorporated; no legal rights were guaranteed; and no allotments were outlawed. Much remained for the incoming Roosevelt administration to consider. It was the Meriam Report, however, that helped start the push toward an Indian New Deal.[54]

■   ■   ■

The first two decades of the twentieth century brought native peoples to the precipice. Religious practices were banned; Indian lands were sold; and tribal trust funds were looted. Health care was nonexistent; food was scarce; and tribal populations decreased. In some ways, Native American survival depended on legal survival, and for legal change to occur, the morality of the treatment of Native Americans had to be exposed. This was the major accomplishment of the Meriam Report.

How did Native Americans see this new development? The hope that was so much a part of this change was lost for many Indians. Although the Depression of the 1930s, so devastating to many others, was not seen by Indians as particularly damaging, it prevented needed funds from being assigned to programs in the Indian Service and so the economic and legal hardships of the past continued.

Survival at the end of the 1920s was not a new phenomenon; it had been an everyday occurrence since the dawn of reservations. But there was an urgency about it. James Welch, a Blackfoot novelist and poet, described this feeling in his poem "Surviving":

The day-long cold hard rain drove
like sun through all the cedar sky
we had that late fall. We huddled
close as cows before the bellied stove.
Told stories. Blackbird cleared his mind,
thought of things he'd left behind, spoke:
"Oftentimes, when sun was easy in my bones,
I dreamed of ways to make this land."
We envied eagles easy in their range.
"That thin girl, old cook's kid, stripped naked
for a coke or two and cooked her special stew
round back of the mess tent Sundays."
Sparrows skittered through the black brush.

That night the moon slipped a notch, hung
black for just a second, just long enough
for wet black things to sneak away our cache
of meat. *To stay alive this way, it's hard.* . . .[55]

4

Native Americans and Constitutionalism

The arrival in Washington, D.C., of the administration of Franklin Roosevelt signaled the assent of a new group of American leaders committed to change in many areas of American life. FDR's people immediately recognized that in the legal maze and economic and social disaster surrounding Native Americans, legal, economic, and social changes were essential. The Meriam Report of the Hoover days established that all was not well with American Indians and that the official federal policy of forced acculturation was cruel and barbaric. Even the Hoover administration had found conditions to be appalling. But what was to be done?

Another kind of approach to the "Indian problem" was needed, so said many associated with FDR, and in 1932 for Native Americans that meant an Indian New Deal. This new wave of reform brought with it many different dimensions to an already complex legal situation for Native Americans. The Depression had to be met head-on at the site of the disaster—the reservation—and the Roosevelt administration promised basic economic aid in a variety of forms. Putting Indians to work in government-sponsored programs, however, would not provide long-term relief, and John Collier, the Commissioner of Indian Affairs, resolved to accomplish fundamental change through the Indian Reorganization Act (IRA) of 1934. With its passage and implementation, a new entity was created that would have an important impact on Native Americans and the federal Bill of Rights. This was the IRA-based tribal government.

Change is not ever an easy process, and it was not for Native Americans in the 1930s. New laws and new interpretations came up against a hardened past. Even the primary goal of the Meriam Report, the creation of an Indian claims commission, was not realized until the end of the Indian New Deal and World War II. And there would be danger in this era of reform. As is often the situation in movements for change, the need to achieve some kind of reform becomes so consuming that precious rights are sacrificed in the art of compromise.

Reorganization and Restitution: Witnessing the Evolution of Tribal Governments as New Legal Variables, 1929–1940

The reformers who came into power in the 1930s were sure that they had found solutions that would be beneficial to Native Americans. Much needed to be done. First, laws were immediately required to stem the erosion of the land base. The allotment system permitted Indians to be stripped of most of their treaty lands. Landless or without land conducive to agriculture, Native Americans gave new meaning to homelessness.

Other matters needed correcting as well. The assault on Indian culture, many reformers argued, must stop. Indian culture should be understood and respected, and if there were parts that might not be useful, only after a period of observation should they be changed. Tribal governments had to be resurrected and given the tools to achieve economic growth, political stability, and legal power. Legal disputes involving Native Americans had to be resolved, from the petty annoyances found in reservation life to treaty wrongs.

Gone were the missionary concepts that had guided the Bureau of Indian Affairs from the 1870s to the 1920s; instead, trained, efficient professionals—anthropologists—were in vogue. In some ways, when it came to Indian affairs, the hallmarks of the Progressive era during Theodore Roosevelt's administration in the early 1900s had been deferred to Franklin Roosevelt's administration in the 1930s and 1940s.

The Depression and Native Americans

The Meriam Report alerted the incoming administration to Native Americans' many needs, and John Collier took many senators on revealing tours of reservations. Even so, FDR's brigade was not prepared for what they found; they were aghast.[1]

The death rate in the United States for all persons in 1926 was 11.8 per 1,000, but for Native Americans, the death rate was more than double the national average, 25.6 per 1,000. On one reservation in Arizona, the Pimas died at a rate of 54 per 1,000. Many diseases not common among the general population of the United States were reaching epidemic proportions on reservations, including trachoma, syphilis, and the prime killer, tuberculosis. The death rate in the United States from tuberculosis was 0.87 per 1,000 in 1926, but for Indians, tuberculosis killed 16.3 per 1,000 in that year.

Indian children were particularly vulnerable. With minimal medical help, Native Americans under the age of one died at a rate of 26.2 per 1,000, whereas Americans in general died at 13.6 per 1,000, nearly half that rate. Indian children dying under the age of three accounted for 36.9 per 1,000 deaths; a comparable age group for everyone in the United States accounted for 16.2 per 1,000 deaths. Guardianship for Native American children meant a certain early death sentence.[2]

Many of the deaths of native people during the Depression were directly related to a lack of income and the grinding poverty of reservation life. Indians owned little property. In surveys conducted in 1930, 42 percent of all Native Americans held tribally owned property, real and personal, which was valued at less than $1,000, and 63 percent of all Indians owned individual property worth less than $1,000.[3]

Income per capita was extremely low, and the Depression made it worse. The data for this period are incomplete, but some Indian nations were surveyed. In South Dakota in 1926, for example, Sioux annual income was reported at $166 per capita, a figure that fell significantly as drought and the Depression continued. By 1935, the annual income on Sioux reservations had dropped to $67 per capita, with less than 10 percent coming from farming or ranching. This is not surprising, as Native Americans in South Dakota and elsewhere simply had no means to earn a living. Although they were supposed to farm, they had virtually no tools other than an occasional rake and hoe, and the lack of water made the land virtually useless for agriculture.[4]

Surveys of specific reservations conducted in 1930 showed a desperate situation. Of the sixty-four tribes examined, sixty-two registered an average individual income of below $500 a year. The largest Indian nation, the Navajos, earned $135 per capita a year. The income range extended from $13 per capita a year on the Carson Reservation in Nevada, $27 for the Silets in Oregon, and $28 for the Catawbas in South Carolina, to $11,265 for the Osage.[5] The Osage benefited from oil and gas revenues, but that would quickly change.

Annual income figures must be contrasted with federal appropriations to the Bureau of Indian Affairs (BIA) and its mismanagement. In 1930, $30 million was appropriated to the BIA to administer the guardianship of 250,000 Native Americans on reservations. This constituted approximately $120 per Indian, more than many made in income for a decade. Over 5,300 employees worked for the BIA, or roughly 1 BIA employee for 43 Indians. Tribal funds provided additional resources for funding the bureau, but regular accounting of these funds was frequently ignored. Estimates of corruption rose to as high as $100 million lost from tribal funds since 1880. On the Blackfeet Reservation in Montana alone, $4.8 million was unaccounted for by the BIA.[6] Thus the guardianship of Native Americans was characterized by greed as well as death. As Oliver La Farge wrote in 1934, "Our own greed is choking us."[7]

Conditions for Native Americans were further complicated by allotment. Many Indians lost their lands. For example, in Oklahoma, among the Comanches and Kiowas, 6 of every 10 were landless, and 7.2 of every 10 Cherokees, Choctaws, Seminoles, Chickasaws, and Creeks also were landless. Oklahoma tribes lost 13.5 million of their 15 million acres because of allotment.[8] It was estimated that by 1934 in the United States, 100,000 Native Americans were landless, homeless, and mired in poverty. As many as 50 percent of all Indians did not own land, and on the reservations less than 10 percent of its residents had anything approaching adequate housing.[9] With no income, no land, no medical care, and no home, Native Americans were destitute.

The Indian New Deal

John Collier, the new Commissioner of Indian Affairs, wanted to change the entire direction of federal policy toward Native Americans. He sought to establish a new policy of "conservation"—conserving Indian resources and cultures—a policy with important Bill of Rights ramifications.[10] Toward that end he dedicated his life, whether or not others, including Native Americans, wanted these changes.

Just before Collier began, Congress had been restless. Law and order on the reservations once again took up its time, and in 1932, Congress decided that the Major Crimes Act of 1885 needed to be amended. Accordingly, it raised from seven to ten the number of crimes committed on reservations having exclusive federal jurisdiction, by adding incest, assault with a dangerous weapon, and robbery to the list.[11]

By 1933 when Collier assumed office, his foremost task, as D'Arcy McNickle, a Montana Blackfoot, remembered, was how best to save the Indians.[12] Collier's tenure at the BIA at first might have looked tentative. In the hundred years of the BIA's existence, there had been thirty-three commissioners, and only three had lasted longer than seven years. Few of them had ever ventured west of the Mississippi River, although Collier's empire included more than 5,000 employees administering 200 tribes who spoke over 50 languages. Few, if any, BIA employees spoke any Indian language when John Collier took over the commissioner's office. Indian nations resided on 84 separate reservations and agencies and sent their children to 65 boarding schools and 128 day schools under the administrative arm of the commissioner.[13] Collier might achieve the power to make changes if he lasted, and he stayed for twelve years. In order to help him, FDR promptly abolished the Board of Indian Commissioners, which strongly favored continuing the forced acculturation and allotment policies.

John Collier immediately set about to make a number of changes in the way that the federal government treated Native Americans, particularly in regard to abrogating basic constitutional rights pertaining to religion and culture. That is, religious ceremonies were forbidden on reservations; Indians were forced to join Christian religions and attend churches; children were shipped off to Christian boarding schools; Indian languages were forbidden; and hair was cut under penalty of lost rations. Collier's response was to issue, in early January 1934, a departmental order, Circular No. 2970: "Indian Religious Freedom and Indian Culture."[14]

Collier's order for cultural and religious freedom demanded full constitutional protections for Indian religious thinking. He decided that even though the courts had not applied the First Amendment to Native Americans, he would apply it by administrative action. In part, the memo stated,

No interference with Indian religious life will be hereafter tolerated. The cultural history of Indians is in all respects to be considered equal to that of any non-Indian group. And it is desirable that Indians be bilingual—fluent and literate in English, and fluent in their vital, beautiful, and efficient native languages.[15]

This first act set a tone that the Indian New Deal attempted to follow. Collier's next order generated a great deal of resistance. On January 15, 1934, Collier prohibited compulsory attendance by Indian children at religious services in boarding schools without Indian parents' permission. Protestant and Catholic missionaries alike castigated Collier, branding him a devil worshiper and a supporter of peyote. Collier saw the practice simply as violating the spirit of the Bill of Rights.[16]

Although Collier made several important decisions in his first months in office, he found that there were limits to his power. His executive orders did not have the permanent force of law; if he were to leave, his legacy could easily be undone. Moreover, there were hundreds of other regulations promulgated by previous commissioners that restricted Native American rights, and they were constantly being thrown up as impediments to what Collier perceived to be the beginnings of an Indian New Deal. He therefore turned to Congress and first successfully sought to repeal twelve espionage or gag rules that restricted Native American civil liberties. Congress approved the abolition of the Espionage Acts that had limited freedom of speech by Indians and persons communicating with Indians, allowed the Commissioner of Indian Affairs to remove from reservations persons considered detrimental, endorsed military maneuvers on reservations, and required passports for Indians who wished to travel from one reservation to another.[17]

Related to the repeal of the Espionage Acts was the repeal of departmental regulations on law and order. The superintendent of each Indian reservation appointed judges to the Courts of Indian Offenses, and each judge acted as a prosecuting attorney, justice, jury, constable, and jailer. Abuses were prevalent, because the superintendents could remove the judges at will. Collier sought to stop the abuses by the Courts of Indian Offenses by issuing a new set of law-and-order guidelines on November 27, 1935.

Officials in the BIA were ordered to stop obstructing Indian courts. Removing and hiring Indian court judges now required approval by the Indians residing on the reservation. For the first time, Native American defendants were allowed to summon witnesses to trial proceedings, to raise bail in order to be freed from incarceration, to see formal charges of accusations, and to request a jury trial. These fundamental rights of the First, Sixth, and Eighth Amendments were now extended to Courts of Indian Offenses by executive order. This new series of regulations also replaced the Code of Indian Offenses, a list of rules used primarily to attack Indian cultural practices. A new code of misdemeanor offenses was to be formulated by superintendents in consultation with Native American reservation residents.[18]

Still a fundamental problem remaining for all Native Americans, John Collier and his followers believed, was the allotment system, which separated Indians from their lands. Once he became Indian commissioner, Collier immediately ordered a moratorium on the sale of tribal land, and he stopped issuing certificates of competency that allowed individual Indians to sell their allotments. Of course, these actions were all that Collier could do administratively until Congress abolished allotment.[19]

The initial steps taken in the Indian New Deal were a breakthrough. Collier used his office to try to restore a sense of Indian community, to protect Indian culture, and to institute some semblance of legal rights for Native Americans. But much remained to be done, and Collier pegged his hopes on what was planned to be the cornerstone legislation—the Indian Reorganization Act (IRA).

The Indian Reorganization Act of 1934

On the twentieth anniversary of the Indian Reorganization Act, John Collier reflected on its purposes. "That intention," said Collier, "was that the *grouphood* of Indians, twenty thousand years old in our Hemisphere, should be acknowledged as being the human and socially dynamic essential, the eternal essential, now and into the future as of old." The IRA, according to Collier, restored cultural determination for Native Americans, "this fundamental of mental and moral health to the only groups which official[ly] or governmentally had been denied it, the Indians."[20]

Collier and other New Dealers wanted legislation to restore what he called the "bilateral relationship" between Indians and the United States. He correctly believed that bilateralism had been the original basis of United States–Native American legal relationships. However, from the 1880s until the 1930s, federal governmental fiat had replaced this shared governance, with horrendous consequences. The Bureau of Indian Affairs developed an authoritarian structure of a monolithic character, with the General Allotment Act signaling the beginning of this process. Collier decided that it must be destroyed before worthwhile change could be ensured.

With these and other concerns in mind, Collier and his associates drew up a forty-eight-page, four-part draft document for Congress to consider. Title I of Collier's proposed Indian Reorganization Act was called "Indian Self-Government," according to which each tribe could petition for a home-rule election, and if approved, the tribe could then charter a tax-exempt corporation that would be authorized to set up businesses, manage tribal property, and borrow funds from the federal government for economic development. "Special Education for Indians," Title II, was designed to promote Indian civilization, including Native American traditions in the arts, and subsidize primary and secondary education of Native Americans. Title III, "Indian Lands," abolished allotment, restored tribal ownership of "surplus" lands, restricted alienation of allotted land, and provided tribes with funds to buy additional land to reconstitute the lost reservation base. The last section, Title IV, "Court of Indian Affairs," established a new federal court comparable to a federal district court to hear disputes involving Native Americans.[21]

This initial version of what became the Indian Reorganization Act was not written by Indians or with consultation with Native Americans. In this sense, it was not a departure from tradition. It also did not contain what had been regarded by many as the single most important need for the recognition of Native American civil liberties, an Indian claims commission. In any case, the Indian Reorganization Act was sponsored by Edgar Howard of Nebraska in the House of Representatives and Burton K. Wheeler of Montana in the Senate. It quickly became known as the Wheeler–Howard bill.

John Collier realized that he needed the Indians' consent to such a radical change in their lives, and so he set out to tour reservations throughout the country to explain the Wheeler–Howard bill. Indeed, he believed in bilateralism as a legal concept, even if it came at the end of the process. These "reservation meetings" were well attended, and, much to his surprise, there was significant opposition to the bill.

Navajos told the New Dealers that their tribal council was fine just as it was. The members of the San Juan Pueblo liked the blend of democracy and absolutism in their government. Comanches were concerned about their loss of inheritance and allotment lands. The Eastern Band of Cherokees objected because they were happy with their own North Carolina–based corporation, which was sanctioned under state law. Many Oklahoma tribes, such as the Quapaws, objected to Collier's proposal because they feared a return to the "old ways"; they had become Christianized and thought that the proposed bill might bring back traditional Indian religions.[22]

Other Native American opponents saw it differently. To agree to the IRA, they reasoned, was to give up too much sovereignty. They argued that the provisions for tribal government diminished their inherent powers of self-government. American courts had held since John Marshall's day, and most recently in *Talton v. Mayes* (1896), that the only way that Indians could give up their self-government sovereignty was by consent, through either a treaty or legislation. Others saw the Wheeler–Howard bill as attacking their rights to future claims based on treaty violations, and they objected to the provision requiring one-quarter or more Native American blood in order to be a member of a new tribal community organized under the proposed bill. Tribal membership, they believed, should be determined by the particular Indian nation involved and thus constituted a sovereign right. Frank Ducheneaux, leader of the Cheyenne River Sioux, felt that the Wheeler–Howard bill "did not relieve the Indians from the control of Congress and the Secretary of the Interior, but limited their sovereign rights which had never been done before formally."[23]

Non-Indians objected as well. Missionaries complained that Indians were being segregated, that Indian lands should be taxed, that Indian courts would allow Indians to escape punishment, and that Christian spiritual values had to be strengthened, not attacked. Cattlemen and mining companies worried about their "investments." Assimilationists such as Flora Seymour, a former member of the Board of Indian Commissioners (which had been abolished), accused Collier and others of being communists and favoring paganism. She argued that the Bill of Rights purposely did not apply to Indians because they needed to change their soviet-like society. Ludicrous as these arguments sound, they were taken seriously by many members of Congress and the non-Indian American public.[24]

The one Native American leader who perhaps understood the controversy best was Joseph R. Garry, a Coeur d'Alene Indian from Idaho and president of the National Congress of American Indians. He observed four distinct reactions by Indians to the IRA: Two attitudes were pro-IRA, and two led to opposition. Most tribes, according to Garry, did not understand the implications of each phase and so opposed the Wheeler–Howard bill because of a lack of information. Others who did understand it and believed that they had accomplished a great deal without governmental help saw little value in it. Other Indian nations were more amenable to the bill. Small tribes who had lost most of their lands readily embraced the new law, and some tribes, both large and small, because they had had so many unsatisfactory dealings with the federal government, simply accepted the new ideas without discussion.[25]

Despite the uproar, Collier, Wheeler, and Howard pushed ahead but made considerable modifications before the bill became the Indian Reorganization Act. Collier proposed a compromise that eliminated many of the controversial points, but kept what he felt was essential. Senator Burton Wheeler introduced the new bill to the Senate on May 18, 1934. The Wheeler–Howard proposal was reduced from forty-eight pages to eleven, and by the time it was approved in both houses, the Indian Reorganization Act had nineteen brief sections in six pages.[26]

What did the Indian Reorganization Act do? It abolished allotments. It continued in perpetuity the trust periods and alienation of lands, unless Congress wished to change their status. ''Surplus'' lands were restored to tribal ownership; the sale of restricted Indian assets was forbidden except to members of the tribe; and the Secretary of the Interior could purchase land, water rights, and surface mineral rights to add to the Indian land base. Congress, if it desired, could authorize up to $250,000 annually to pay for tuition for Native Americans to attend vocational and trade schools, and the IRA suggested placing $10 million in a revolving fund for the Secretary of the Interior to lend to Indian-chartered corporations to promote economic development.

Lobbyists succeeded in restricting the application of the IRA to certain tribes. Excluded from important portions of the act were Indians residing in Oklahoma, who constituted approximately one-half of all Native Americans, and those residing in Alaska. In 1936, many of the IRA's provisions were extended to them in separate laws, the Oklahoma Indian Welfare Act and the Alaska Indian Reorganization Act.[27]

Two important parts of the original bill's Indian self-government section were modified and retained. Under the approved IRA, Native

Americans residing on reservations had the right to organize tribal governments by adopting an appropriate constitution. Such a constitution gave tribal governments the right to employ legal counsel, to prevent the sale or lease of tribal lands or assets, and to negotiate with federal, state, and local governments. The Secretary of the Interior, however, still had to approve the choice of legal counsel and legal fees charged and advise tribes on appropriations and federal projects. The act excluded approximately 40 percent of all Indians in the United States from adopting tribal governments under it, and eligible tribes could participate in this exercise of constitutionalism if they approved the IRA. Elections were to be held on all reservations as called by the Secretary of the Interior. If a majority of all eligible voters rejected the IRA in an election in which a minimum of 30 percent of the eligible voters participated and that was held within three months of the IRA's passage, then the tribe was not to be covered by the IRA and so could not adopt a constitution.[28]

If a group of Native Americans approved the IRA in a Department of the Interior–sanctioned election and proposed and received approval for a constitution, they could petition for a charter of incorporation that would enable their Indian nation to borrow investment capital from the federal economic development revolving fund. The charter had to be approved by the Secretary of the Interior and then favorably voted on by the tribe. Charters allowed the tribe to purchase, manage, or dispose of real and personal property, except for land held in a reservation for a period of ten years from the signing of the IRA. An incorporated Indian nation could also conduct any form of corporate business consistent with federal law. Charters could be revoked only by a specific act of Congress.[29]

Thus the legislative process significantly altered Collier's reforms. The basic purposes of the Indian New Deal were deleted from the preamble. Portions of those sections promoting Indian culture were left out. Mandatory provisions concerning the assumption under tribal ownership of previously granted Indian allotments were changed to a voluntary action. No self-government or economic aid could go to nonreservation Indians or to lapsed members of individual tribes. The Secretary of the Interior's powers over Indian tribes were enhanced, because the inherent sovereign rights of Indian governments would be compromised if a tribe agreed to the IRA's provisions and permission had to be obtained before constitutions and corporate charters could go into effect. Finally, Title IV of the original bill, setting up a new federal court, was eliminated. Even so, Collier was pleased with the Indian Reorganization Act. It proved to be the single most important federal legislation ever passed directly relating to Native Americans. What was left out he would attempt to implement through administrative acts and other New Deal programs.

Two fundamental principles were established in the IRA. First, John Collier and the Indian New Dealers wanted allotment to be abolished, and this was achieved. Second, they wanted the BIA to work with Indian nations with what Collier called "indirect administration." Collier urged less paternalism so that Indians could retain and build a strong cultural presence and work within the American system. Collier called this the creation of a Red Atlantis. This second principle was also a part of the IRA, but its application proved to be controversial among Native Americans. "Indirect administration" still prevented the exercise of basic rights, and less paternalism was not the end of paternalism.[30]

The Evolution of Constitutions and Tribal Governments

Modern constitutionalism for Native Americans is derived from Sections 16 and 17 of the Indian Reorganization Act, according to which new tribal governments were established and Indian nations were incorporated. The first step in the process was to obtain Native American approval of the Indian Reorganization Act.

The terms set forth under the IRA required that tribes hold an election within one year of the signing of the act; the deadline was June 18, 1935. To defeat the IRA's application, a majority of all adult Indians in a particular tribe on a reservation had to vote against it. Each tribe had one chance to reject the act.

This kind of democracy posed several problems in its application to Native Americans. The enactment provision of the IRA weighted the referendum process in favor of approval, and so those who did not vote were counted as if they had voted affirmatively. Many tribes traditionally did not consult all adult members in reaching tribal decisions, and many remembered traditional forms of governance. Tribes that were matrilineal, such as the members of the Iroquois Confederation, valued the opinions of women more than those of men. Moreover, the concept of consensus was often much more of a cultural necessity for an Indian nation's decision making than was the concept of majority rule. Still, the Commissioner of Indian Affairs ordered the elections to be held.

There has been a great deal of confusion surrounding the ratification results of these first Indian elections. Historians have been unable to verify accurately each election. But the most comprehensive examination shows that 252 Indian tribes and bands voted on the IRA referendum. Of the 252, 174 voted in favor of the act and 78 voted against it. Thirteen other Indian bands listed on the BIA election roll as eligible to vote did not participate because they no longer contained any members.

Approximately 97,000 Indians were eligible to vote. About 38,000, or 39 percent, voted yes; about 24,000, or 25 percent, voted no; and about 35,000, or 36 percent, did not vote.[31] However, the 36 percent who did not vote were counted as having voted yes. The idea that not voting meant an affirmative vote for the IRA caused a tremendous uproar and created divisions on some reservations. Congress eventually changed the rule to a majority of those actually voting, but this ratification change did little to alter what had already occurred, since the amendment to the IRA was passed on June 15, 1935, only three days before the referendum deadline.[32]

The majority that voted no to the IRA but still were forced to live under the act included seventeen separate tribes. For example, at the Santa Ysabel Reservation near Riverside, California, forty-seven tribal members voted no, fourteen voted yes, but sixty-one refused to vote. Thus the Department of the Interior counted the vote as seventy-five yes to forty-seven no and accordingly placed the Santa Ysabel California Mission Indians under the IRA.[33]

Of the 174 tribes that ratified the IRA, 92 nations wrote and approved constitutions. Seventy-two tribes either did not submit constitutions for the Department of the Interior's approval or voted them down in a constitutional election. Of the 92 constitutionally based tribal governments that were created under the IRA, 71 nations formed corporations to take advantage of federal funds set aside for economic development.[34] Thus only 37 percent of the tribes and bands eligible to approve the IRA and constitutions did so, and only 28 percent approved all three aspects of the constitutionalism package.

Indians living in Oklahoma and Alaska initially were not covered by the IRA, but they were granted coverage as a result of congressional action in 1936. Under the Oklahoma Indian Welfare Act (OIWA), all Oklahoma Indians except the Osage were automatically placed under the IRA umbrella. They did not have the option of approval or rejection of the OIWA. Tribal governments and tribal incorporation charters were authorized through a constitution-approval procedure, and a $2 million revolving credit fund was set up for Oklahoma Indians. By 1940, eighteen Oklahoma tribes had ratified constitutions, and of those eighteen, thirteen had approved charters of incorporation.[35]

Alaska's Indians also were treated specially. In 1936, the Alaska Indian Reorganization Act (AIRA) was approved. Under this direct extension of the IRA, Alaska Native Americans were entitled to set up reservations, buy land, approve tribal governments, and incorporate. Like Oklahoma Indians, they had no choice but to come under the AIRA, but unlike Oklahoma Native Americans, they received no special credit fund. Alaska's Indians competed with all others for the IRA's economic development money. Because reservation status was not given to most Alaska Indians, this represented a significant change for them. Native villages were recognized as reservations. Under the AIRA, forty-nine Alaska Indian communities had adopted constitutions and charters by 1940.[36]

Those tribes that participated in the IRA were treated as "organized governments," and those that did not were termed "unorganized governments." The Department of the Interior institutionalized discriminatory practices against those tribes that did not participate in the IRA. Although organized governments had more ready access to BIA programs and funds, several important tribes still refused to participate. Most noticeable were the Navajos, the largest Indian nation in the United States.

For many years, the Navajos governed by leadership consensus in a council based on traditional divisions within Navajo society. In 1923, the Secretary of the Interior dissolved the traditional institutional forms of Navajo tribal leadership and refused to work with it. He then authorized a council to be formed, and this new illegitimate government gave power of attorney to the secretary, who authorized long-term oil leases on the Navajo Reservation for $1,000. The leaseholder then sold rights to the oil for $3 million.[37]

It was not surprising, then, that when John Collier visited the Navajos, they were suspicious of the Department of the Interior. Although their traditional leadership continued to function, it did so against great odds. When the IRA election was held, the Navajos, with a tribal population listed at 43,135 and 15,900 eligible voters, cast 7,992 votes against the IRA and 7,608 for it. What was amazing was that on the largest reservation in the United States, reaching into three states and containing few passable roads, only 300 Navajos who were eligible to vote did not. Thus the final tally provided an 84-vote victory for the opponents of the IRA. Collier was furious with the Navajos, and he threatened them. Later he allowed Department of the Interior authorities to disrupt Navajo society with forced reductions in sheep and goat herds on the Navajo Reservation.[38]

The Navajos eventually were allowed to participate in a revolving credit fund. In the Navajo–Hopi Rehabilitation Act of 1950, a special $5 million economic development package was devised for the Navajos. Tribal councils were to apply for credit from the fund for special projects, and the Hopi and Navajo traditional tribal councils were allowed to participate in the fullest possible way in administering their tribes. In effect, Congress recognized the traditional governments of the Navajo and Hopi.[39]

Before the IRA was enacted, thirteen tribes already had working constitutions. Of the thirteen, four Indian nations voted to organize under the IRA, but decided to keep their existing constitutions. These included the Eastern Band of Cherokees in North Carolina, whose constitution dated to 1897; the Menominees of Wisconsin, with a 1928 constitution; the Red Lake Band of Chippewas in Minnesota, whose constitution dated to 1918; and the Standing Rock Sioux of North Dakota, with a 1914 constitution. Nine tribes, including the Navajos, had constitutions that had been drawn up before 1934, and they refused to approve the Indian Reorganization Act. Five other Indian nations adopted constitutions after the IRA was implemented; they also refused to approve the Indian Reorganization Act. Obviously, constitution making was not always tied to the IRA, nor was it a new undertaking for all Native Americans.[40]

Some tribes drafted their documents carefully and thought long about the constitutional ramifications of each section. One such tribe was the Cheyenne River Sioux in South Dakota. Although there had originally been opposition to the IRA on the Cheyenne River Sioux Reservation, there was a base of support as well. When the vote was taken on October 27, 1934, of 1,420 eligible voters, 653 voted yes, 459 voted no, and 308 did not vote. A constitution was approved on December 27, 1935, but a corporate charter was never ratified.[41]

The preamble to the Cheyenne River Sioux Constitution states that the purpose of their tribal government is "to conserve our tribal property, to develop our common resources, to establish justice and to promote the welfare of ourselves and our descendants.''[42] The constitution contains nine articles, pertaining to territory, membership, governing body (the tribal council), powers of self-government, elections and nominations, removal from office, referendum, land, and amendments. Several legal issues are central to this constitution. First, the Cheyenne River Sioux refused to include provisions for the Secretary of the Interior's approval of the tribal council's powers. According to Frank Ducheneaux, who helped draft the Cheyenne River Sioux Constitution, there was no question that the Sioux retained the inherent sovereign powers of governance. Ducheneaux remembers that when the constitution was drafted, he

> pointed out that the I.R.A. did not grant us any rights that we did not already have as a sovereign nation and that by voting to accept the act and by drafting a constitution in line with Section 16 of the Act, the Tribe was giving their consent to having our authority limited by this Act.[43]

The constitution reflects this viewpoint in its final form.

The Cheyenne River Sioux did approve, however, some powers for the Secretary of the Interior. According to Article IV, Section 1, the Secretary of the Interior has to agree to condemnation proceedings (Subsection l), to ordinances regarding the inheritance of property (Subsection n), and to the appointment of guardians for minors and mental incompetents (Subsection o) in order for these specific kinds of ordinances passed by the tribal council to become law. Article IV, Section 2, of the constitution carefully outlines how and when the Secretary of the Interior can review ordinances passed by the tribal council. The secretary can veto only those laws concerning the few matters specifically delineated in the constitution. The council retains a significant reservoir of legislative, executive, and judicial power.[44]

Of significance is the absence of a ''bill of rights'' in the Cheyenne River Sioux Constitution. Some of the other tribes' constitutions do have a ''bill of rights'' provision. In addition, certain of the Cheyenne River Sioux Council's provisions infringe on some rights granted in the United States Constitution. For example, in Article IV, Section 1, of the Cheyenne River Sioux Constitution, the council is given the power to require individual members of the tribe or other residents of the reservation to assist with community labor (Subsection h). A poll tax is to be charged to all voters, although community labor can be used to satisfy the tax requirement (Subsection i). The council also can remove or exclude from the Cheyenne River Sioux Reservation any nontribal member who is deemed morally or criminally injurious to tribal members (Subsection j).[45]

The Cheyenne River Sioux Constitution preserves as many rights as possible and yet stays within the guidelines offered by the IRA and the Department of the Interior,[46] but other tribes were not nearly as fortunate. The San Carlos Apaches of Arizona adopted a constitution drafted by their own reservation superintendent, which granted a great deal of power to the superintendent. This arrangement worked well as long as the person holding the position was responsive to tribal wishes, but in November 1938, a man named McCray took over and began systematically looting tribal funds without the tribal council's permission. According to Clarence Wesley, chairman of the San Carlos Apache Tribal Council in the 1950s, when "members of the Council would complain and object . . . Supt. McCray would say, 'I am the boss.'" The tribal office was open for only one hour a day. Finally, McCray died, and his replacement began the formal operation of the tribal council and the San Carlos Apache constitution.[47]

In South Dakota, the Indian Organization Unit, a task force organized by Collier to implement the IRA, sent anthropologists and lawyers into the field to work with the Sioux. This action caused much debate and friction, creating a number of tribal factions. The ensuing constitutional battles have been described as pitting the "New Dealers"—the younger, educated, mixed-blood Sioux—against the "Old Dealers"—the older full bloods who already held tribal power. Of the nine Sioux reservations or agencies in South Dakota, eight approved the IRA; of the eight IRA-authorized Sioux organizations, six of them ratified constitutions; and of the six Sioux constitutional governments, three established tribal corporations.

Political hardball was played on the Sioux reservations. At Rosebud, the tribal council opposed the IRA because the Oglala objected to campaigning for office. They believed that leadership was earned by merit, individual work, and respect. Those who spoke out were branded as "agitators" by reservation agents, and Collier took steps to "isolate them."[48] At Cheyenne River, when Sioux leaders, including Clyde Low Dog, wanted to see Collier to complain about the process, the superintendent, Walter F. Dickens, wrote ahead to prepare Collier for their visit. Dickens described Clyde Low Dog to Collier as an "underhanded, vicious character, and his name is the most descriptive of him that could be made."⁴⁹

Fortunately for the Rosebud Sioux, their constitution was adapted to provide representation based on local dance halls that conformed to tribal clans. Pine Ridge was not so lucky, because the constitution recommended by the Indian Organization Unit and adopted there based tribal council representation on farm districts that had no relationship to traditional Oglala government practices. The constitution of the Omahas in Nebraska allowed for seven council positions elected at large. However, the anthropologist who formulated the constitution, basing representation on the seven clans traditionally involved in tribal decision making, forgot that the clans were not equal in electoral strength, and because the majority ruled, one or two usually dominated the council elections. Thus on the Omaha and Pine Ridge reservations, political disputes disrupted the nascent forms of constitutionalism.⁵⁰

■   ■   ■

By 1940, the vital life signs of Native Americans showed some improvement. The Depression and the assimilationist policies of the BIA, so devastating in concert, had not succeeded in destroying their resilient culture, but life still was very difficult. Some of the Indian New Deal measures helped immensely on the reservations. More and more Indians were gaining a higher income and better housing. The Indian land grab stopped. Death rates started to drop, and health services made inroads into Indian diseases. Reservations stabilized.

Tribal governments were also feeling their way. Some of the tribes benefited almost immediately by passing constitutions and incorporating. Rudimentary constitutionalism took hold, if only because the federal government under John Collier's leadership virtually refused to recognize any other form of legal relationship with the tribes. Tribal investments, not the BIA, helped bring jobs to the reservation for the first time, although these improvements were still very fragile and could not afford an interruption.

The Indian Reorganization Act provided great hope for many Native Americans. Legalists have termed it the single most important piece of Indian law ever implemented in the United States. The circumstances surrounding its adoption necessitated compromise, and the competing interests turned loose by the IRA made its success a tenuous matter as well. Indeed, almost as soon as the IRA was in place, opponents emerged calling for its abolition.

The Indian New Deal and the Indian Reorganization Act engendered in Native Americans a feeling of cautious resolve. No doubt many Indians may have been pondering the same message found in a Pawnee poem, "Is This Real":

> Let us see, is this real,
> Let us see, is this real,
> This life I am living?
> You, Gods, who dwell everywhere,
> Let us see, is this real,
> This life I am living?[51]

5

Dualities of Indian Law

The 1940s marked a legal crossroads for Native Americans. The Indian New Deal functioned, but it soon abruptly ended amid the prospect of a terrible war. Even Indian lands in Alaska were invaded by the Axis powers. Although many New Deal programs did not survive, the tribal governments did, and they asserted their power and independence.

This legal revolution on the reservation promised changes that were not always well received. Tribal governments took liberties with Indian individual rights as perceived by those with a bill of rights tradition, and the tribal authorities prevailed. State governments also recognized tribal governments as a threat to the state's traditional way of doing legal business, and the states became aggressive in asserting their powers over Indian lands and persons. Confrontations between tribes and states in the legal arena were much more frequent in the 1940s, and the states often won. Tribal governments embodied a duality of law: They gained much greater legal control over reservations, while they were attacked by the states and sought help from federal lawmakers.

This legal drama joined a movement to resolve Indian claims for past violations of treaties and executive agreements. Again, the legal interplay took on dual characteristics. Tribes were allowed to sue for specific damages, but there were numerous restrictions, and the process was long, complicated, and fraught with potential undesirable ramifications. The creation of the Indian Claims Commission ushered in an era of fairness as a new legal standard, but it did not prevail. Although the

Indian Claims Commission Act represented the last plank of the Indian New Deal, it was implemented in an atmosphere of suspicion and distraction.

Reform and Regulation: Consolidating Indian Rights Before the Horrors of Termination, 1941–1946

The early 1940s were important legal formative years for Native Americans. During this short time, many of the legal issues debated in the next two decades emerged. These disputes centered on three fundamental aspects of the legacy of the Indian New Deal: the role to be played by tribal governments, the place of Indians in the dynamics of American democracy, and the resolution of past violations of treaties and executive agreements.

Tribal governments were put on firm footing by the Indian Reorganization Act. Most of them had grown out of past traditions, and they had managed to overcome some of the more debilitating aspects of BIA paternalism to develop constituencies on reservations. This did not mean that problems of governance disappeared. All governments learn with experience, and most governments do not have a powerful "Big Brother" looking over their shoulder, which was especially true in regard to Bill of Rights matters.

How the tribal governments might assert their powers over the reservations led to active competition with other governmental institutions. The development of new regulatory powers over an Indian land base meant new legal questions demanding resolution. In particular, during this period, taxation and tribal membership preoccupied Native American governments. In the political arena, Indians tried to gain access to power through suffrage, and they frequently lost. They were not alone in this endeavor, as other American minorities also had minimal success in gaining the franchise until the civil rights revolution in the 1960s.

The end of the Indian New Deal, World War II, and the formation of the Indian Claims Commission all fell within this short time. These three important events caught up Native Americans in a whirlwind of changes, not all of which were pleasant. The end of the mixed blessings of the Indian New Deal meant, at the least, a loss of income. World War II brought displacement but some steady employment. The Indian

Claims Commission gave rise to new expectations. It was a time of gathering momentum for the dualities of Indian law.

Tribal Governments and the Bill of Rights

Some tribal governments readily grasped their new constitutional role. Councils passed useful legislation that helped realize some of the goals of the Indian New Deal. In particular, they adopted tribal codes of law and established tribal courts to adjudicate tribal disputes. This lead to a fundamental dilemma: Where did the tribal governments fit in the penumbra of the United States Constitution and the Bill of Rights?

The question of whether tribal governments could violate the Bill of Rights had important ramifications. For example, in the late 1930s, the Rosebud Tribal Council became upset with a petition from Rosebud Oglalas critical of the council. The council's response was to pass ordinances prohibiting anti–tribal council and anti-IRA gatherings by its opponents on the reservation. This restriction of the right to assemble directly threatened individual Sioux rights, but because the Bill of Rights did not apply and the ordinances were allowed, the right to assemble was suspended on the Rosebud Reservation.[1]

A different problem with similar results occurred at the Taos Pueblo in New Mexico, where the tribal government attacked the Native American Church. Tribal government police officer Antonio Mirabel interrupted a peyote ceremony and arrested three leaders and their followers and confiscated their peyote. At the trial in tribal court, Mirabel acted as prosecutor and judge, finding fifteen Pueblos guilty of petty misdemeanors and fining each $100. During the trial, Geronimo Gomez defended the Native American Church and questioned Mirabel's authority to conduct such a proceeding. For his efforts, Gomez was fined $225, but because he had no money, Mirabel confiscated 300 acres of Gomez's irrigated land. Mirabel also imprisoned several leaders, charging them with witchcraft. Despite attempts by John Collier and other federal officials to intervene, the Taos Pueblo Tribal Council prevailed, and First Amendment religious freedoms were curtailed at the Taos Pueblo in the 1930s and 1940s.[2]

During the 1940s, it became increasingly clear that some Indians who wanted to remain members of their tribal communities had to sacrifice civil liberties to established tribal governments. A Native American who is a tribal Indian has a special legal relationship to the Bill of Rights.

This person has a dual status. First, he or she is a member of an Indian nation and is required to live under its auspices as if it were sovereign; second, he or she is still a ward of the federal government, so there are certain areas in which federal law prevails over individual or collective rights.

Talton v. Mayes (1896), decided before the Indian Reorganization Act, established that tribal authorities did not have to guarantee due process to individuals under the Fifth Amendment because tribes were not federal institutions. Their governments were adjudged to be inherently sovereign except when they ceded powers and responsibilities. The problem of the Fifth Amendment was taken up and then avoided by the Supreme Court in 1938 when it decided *United States v. Klamath Indians.*

When the federal government allowed over 110,000 acres reserved by treaty for the Klamaths to be given to the state of Oregon, the Supreme Court held that the United States owed the Klamaths more than $5 million, the value of the lands taken plus interest, not because of the Fifth Amendment but because of an implied promise to pay for essentially the exercise of eminent domain over a guardianship. *United States v. Klamath Indians* was a forerunner of the Indian Claims Commission and suggested that the federal government had to find ways to rectify past wrongs so that certain concepts found in the Bill of Rights could be applied to tribal governments and, possibly, individual Indians.[3] The Fourteenth Amendment also did not apply to tribal governments because it pertains only to state action. Thus any application or "marriage" of the Bill of Rights to tribal governments through the Fourteenth Amendment was not possible.[4]

It had generally been established that Native Americans were subject to limited federal jurisdiction only collectively. Rarely did individual Indians enter the federal legal system unless they committed a particular crime or violated laws outside their reservations or allotments. Internal legal affairs were the concern of tribal governments. This involved the resolution of civil disputes, the exercise of limited criminal jurisdiction over most misdemeanors, the development of tribal inheritance rules, and the regulation of tribal property. Under the Indian New Deal, tribal governments had the right to legislate in all areas ignored by Congress, and it had already been established that general federal legislation did not apply to Indian nations or individual Native Americans.[5]

Because of this new direction for internal Indian law and a lack of knowledge about the myriad specific federal Indian laws, Secretary of the Interior Harold Ickes decided to strengthen the department's ability to handle these new areas. He appointed Felix Cohen as assistant solicitor of the Department of the Interior and assigned him to compile a survey of Indian law. Cohen meticulously collected all materials relating to federal, state, and tribal law concerning Native Americans and produced the *Handbook of Federal Indian Law,* first published by the Department of the Interior in 1941 and continually updated since. This was an important contribution to Indian tribal governments and to understanding the relationship of Native Americans to the Bill of Rights.[6]

Indian tribal courts emerged from tribal governments as important legal institutions that enforced and interpreted tribal codes. By the end of the Indian New Deal, three kinds of courts existed on Indian reservations. Twelve of the Courts of Indian Offenses established in 1882 and run by the BIA still remained and were used primarily on reservations that did not have tribal governments. The nineteen traditional courts— judicial systems in place long before the IRA—asserted sovereign jurisdiction not connected to tribal governments. Southwest Indian nations such as the Hopi and various Pueblos maintained traditional courts. But the great majority of Indian courts, some fifty-three, were tribal courts organized as a result of tribal constitutions and tribal governments.[7]

The tribal courts quickly became a means to express tribal sovereignty. Although the tribal judges varied in their abilities, some of the better-trained justices were successful in expanding their jurisdiction and controlling the practice of law on their reservations. Some tribal courts adopted rules preventing non-Indian attorneys from practicing law in their proceedings unless they were affiliated with Native American attorneys. Others required a particular Indian language, such as Navajo, to be spoken in court, which effectively precluded nontribal lawyers.[8]

The reception to these courts was mixed. The Papagos established a tribal court that combined the expertise of the legislative branch with that of the judicial branch. Three tribal councilmen were selected to form the Papago Tribal Court, which met each Monday to hear disputes and to enforce the tribal code, which was in constant evolution. Two police officers were hired by the Papago Tribal Council and attached to the Papago Tribal Court. This arrangement seemed to work well for the Papagos.[9]

On the Pine Ridge Sioux Reservation, however, the tribal court proved to be quite unlike that of the Papagos, as the Pine Ridge Tribal Court quickly became identified with corrupt reservation agents. It was used only to enforce unpopular decisions, and it was organized with circuits based on six districts that conformed geographically to the earlier hated ration districts of the 1880s rather than to traditional band-settlement patterns. Confrontation between the Pine Ridge Tribal Court and reservation factions was more the rule than the exception.[10] Thus the tribal courts quickly developed their own traditions and so defied generalization.

New Powers in Indian Country: The Power to Regulate

With the implementation of the New Deal in general and the Indian Reorganization Act in particular, Native Americans found that they, like most other Americans, were subject to new forms of governmental regulation as a part of the Roosevelt administration's focus on relief and recovery from the Depression. For Indians, this initially meant contests among three interested parties for the powers to regulate. The federal government, the state governments, and the new tribal governments all had something to gain. In the 1930s and 1940s, a struggle over the new powers to regulate Native Americans centered on two issues: tribal membership and taxation.

Tribes determined their membership based on traditional criteria and their constitutions. Many of the constitutions effectively codified tribal practices. To be an Indian required both some element of Indian blood and the recognition of Indianness by a Native American community. Although the IRA allowed tribes to classify their membership, it cemented the concept of requiring an Indian to be a member of a federally recognized tribe in order to obtain federal benefits and programs. Federal recognition easily accrued from tribal governments and constitutions, although it was not an absolute requirement.

Disputes arose over the maximum amount of non-Indian blood allowed for a person to be classified a Native American. The range was quite wide. For example, the Uintah and Ouray Utes of Utah required over 50 percent general Indian blood with at least 50 percent Ute blood for tribal membership, whereas the Cherokees of Oklahoma simply needed a small fraction of Indian blood to establish membership.[11] At the Santa Clara Pueblo in New Mexico, the tribe established that the children of a Santa Clara Pueblo man who married a non-Indian woman were automatically tribal members, but that the children of a Santa Clara Pueblo woman who married a non–Santa Clara Pueblo man were not. Even if the Pueblo woman married a full-blood Indian, their offspring were not entitled to tribal membership. This traditional patrilineal rule of Pueblo culture was later challenged before the Supreme Court.[12] Codifications of tribal-membership customs became the subject of considerable litigation in later years, and Bill of Rights issues were at the core of these disputes.

The power to tax exercised by tribal governments was another matter that had important Bill of Rights ramifications and immediately became a source of friction among Indian nations, states, and the federal government. It was established during the New Deal era that the federal government could tax Indian income, even if it had been earned on reservation lands or allotments. In *Superintendent of the Five Civilized Tribes v. Commissioner* (1935), the Supreme Court held that income earned by an Indian from restricted allotments could be taxed by the federal government and that this concept was not at odds with the guardianship responsibilities of the federal government.[13] But income tax aside, most sales or products made on reservations are free of federal tax, as the Revenue Act of 1932 excluded native handicrafts from taxation.[14]

States that wanted to tax Indian lands, incomes, and sales were much more restricted, but during the early 1940s they aggressively sought these powers. The federal government, however, tried to protect the nascent tribal governments from state action. In New York, a federal district judge in 1938 nullified a state court's attempt to stop the Senecas from establishing inheritance laws. Four years later, another federal judge declared unconstitutional a New York law that prevented Indians from ending expired land leases with non-Indians.[15]

Whether states could tax Indian incomes, sales, or products had yet to be decided, and the states sought to force the issue. The states could tax non-Indian property and income earned on reservations, but they had to be careful. They could not tax in those areas that the federal government had already indicated were restricted to federal regulation. And there was some confusion where the tribal governments might fit in. For example, the states might not be able to tax in areas preempted by the Indian tribes themselves.

This last clash invited the tribal governments to act, and they did. Many tribal constitutions allowed tribal governments to tax tribal members. Some required the Secretary of the Interior to approve tax ordinances, whereas others did not permit this kind of delegation of power. Poll taxes were common, and modest per capita taxes were charged. Most tribes, however, were reluctant to extract sizable taxes from their members, even though they clearly had the power to do so.[16] One of the earliest tribes to tax tribal members was the Papago, who imposed a 3 percent tax on all livestock sales except from the tribal herd. The funds earned from the tax were to be used to assist the tribe in all other range and livestock activities.[17]

Governments fought, however, over whether Native American nations could tax non-Indians. Tribes attempted to tax commercial sales on Indian reservations, and initially they were successful. Might tribes be able to tax non-Indians who made only infrequent purchases on reservations? Could tribes tax natural resources obtained by non-Indians or corporations through long-term leases on reservations? By the end of the Indian New Deal, these issues had not been resolved.

The Right to Vote

The evolution of tribal governments and the introduction of limited, or indirect, home rule immediately introduced Native Americans to various aspects of democracy. In particular, the right to vote became an issue, and not just on the IRA or tribal constitutions or tribal charters. Could Indians vote in local, state, and federal elections? In many states by 1946 the answer was negative. In 1946, Felix Cohen wrote that "in a democracy suffrage is the most basic civil right, since its exercise is the chief means whereby other rights may be safeguarded. The enfranchisement of the Indians has been a slow and is still an incomplete process."[18] Why had Indians not been allowed to vote, especially after the passage of the Citizenship Act of 1924, which conferred citizenship on them?

State and local governments actively discouraged or prevented Indians from voting. The argument most often used was derived from language in the unamended United States Constitution. The phrase "Indians not taxed" describes, first, how to determine who may be counted for congressional representation and, second, how direct taxes are to be apportioned. "Indians not taxed" was featured in subsequent legislation, including the Civil Rights Act of 1866 and the Fourteenth Amendment.

California, Washington, New Mexico, and Idaho disenfranchised Native Americans through their state constitutions. South Dakota prevented those Indians who retained their tribal membership from voting. Arizona passed a statute denying suffrage to all Native Americans based on the notion that the general terms of guardianship precluded full citizen participation. In 1936, following its attorney general's opinion, Colorado refused to allow Indians to vote because the state's lawyer incorrectly ruled that Native Americans were not classified as United States citizens, and one year later Utah passed a law denying suffrage to Indians who lived on reservations. Some states, such as Arizona, South Dakota, and Idaho, specifically forbade Indians, on the basis of race, from running for and holding public office. Other states, however, such as Oregon, Oklahoma, and New York, expressly permitted Indians to vote and participate in American democracy.[19]

Arizona seemed especially active in denying Indians the right to vote. In 1928, the Arizona Supreme Court held in *Porter v. Hall* that Indians living in Arizona could not vote in local or state elections because they were wards of the federal government. Arizona law stated that those persons "under guardianship" could not vote, an interpretation that disenfranchised 10 percent of the eligible voters of the state. Arizona's attorney general ruled first in 1928 and again in 1944 that Indians who left reservations were also ineligible to vote, even if they paid state taxes and were subject to state laws.[20] The Fifteenth Amendment was inoperative in Arizona and other states.

Because most Indians could not participate in the democratic process, they seemed less inclined to see how useful it might be on their own reservations. Moreover, when opposition to the Indian New Deal became emboldened in the 1940s, they had little means to fight. World War II solidified the inevitable.

World War II and the End of the Indian New Deal

Opponents of the Indian New Deal gained the upper hand in the late 1930s and early 1940s. They wanted to end the IRA, tribal governments, and special New Deal programs for Indians. Bills were introduced in Congress to achieve these goals, and even the sponsor of the Indian Reorganization Act, Senator Burton Wheeler, reversed himself and sided with the IRA opposition. But the courts would not defeat the Indian New Deal, and Congress proved reluctant to act. Nonetheless, another, and perhaps more effective, way was found to stop the Indian

New Deal. Massive cuts in funds for Indian New Deal agencies proved politically popular.

After 1938, the House of Representatives annually attempted to stop appropriations altogether for the IRA. Although the funds were never actually canceled, congressional appropriations were far short of IRA authorizations. Under the IRA, up to $250,000 was allowed to be spent annually on the Indian Organization Unit, which was assisting tribes in setting up tribal governments and corporations, but only $150,000 was appropriated. The $2 million for land purchases dropped to $1 million, and the revolving credit fund fell from $5 million to $2.5 million. Education loans to be set at $250,000 a year were never above $175,000.[21]

There also were many problems with the implementation of the IRA. Corruption still existed in Indian education, and funds for housing were misused. Some states, such as Arizona, openly discriminated again Indians in the administration of Social Security. Those tribes that questioned BIA authority, especially in regard to approving the IRA or tribal constitutions, were penalized by the BIA. They simply did not receive information about programs, or they were denied services that were given to more cooperative Indians.[22] Probably the biggest mistakes of the Indian New Deal involved the largest Indian nation, the Navajos. Although the Navajos were not under the IRA, in 1933 John Collier ordered them to kill 400,000 sheep, goats, and horses in order to stop erosion on their vast reservation. Small herds were wiped out, and many Navajos became destitute.[23]

Indian New Deal opponents and program mismanagement did not cause the Indian New Deal to end. Instead, it was World War II that effectively stopped it. The impact was overwhelming in every respect. Collier, who resigned after serving the longest term ever as Commissioner of Indian Affairs, witnessed the strangulation of his programs by a greater need.

The BIA lost a large number of personnel. Doctors, nurses, and dentists transferred from Indian health services to the war effort. Interagency cooperation over Indian New Deal programs ended with the focus on the war. Irrigation programs were canceled; conservation programs were ended; land purchases were put on hold. The revolving credit fund for economic investment was frozen in 1942. With no forms of federal employment available and with training programs curtailed, 50 percent of the able-bodied Indian men left their reservations to enlist or to find work in war industries, thereby destroying the progress in Indian self-government and cooperative economic programs. After the war, the same Indians returned to the reservation to unemployment.[24] Indeed, World War II profoundly changed Indian life. Those Native Americans left on the reservations lived under pre–Indian New Deal conditions, but the end of the war also ushered in a new event never before experienced by Native Americans: The Meriam Report's recommendation was finally realized with the creation of the Indian Claims Commission (ICC) in 1946.

The Indian Claims Commission

Section 15 of the Indian Reorganization Act of 1934 anticipated the last gasp of the Indian New Deal, stating: ''Nothing in this Act shall be construed to impair or prejudice any claim or suit of any Indian tribe against the United States.'' Tribes were also protected from benefits they received under the IRA being held against their claims. Section 15 further elaborated, ''It is hereby declared to be the intent of Congress that no expenditures for the benefit of Indians made out of appropriations authorized by this Act shall be considered as offsets in any suit brought to recover upon any claim of such Indians against the United States.''[25] The IRA was a prelude to the Indian Claims Commission Act of 1946.[26] John Collier thought of them as related, if not continuous, pieces of legislation.

A number of factors influenced the creation and passage of the Indian Claims Commission Act.[27] Persistence on the part of New Dealers was important. A pro-IRA pamphlet, edited by Jay B. Nash in 1938, gave an example of the strong feelings held by Indian New Deal advocates. It unabashedly called for passage of the act. Wrote Nash, ''The honor of the United States is still compromised by hundreds of broken treaties with Indian tribes. . . . The responsibility of removing this blot upon our national honor still rests with Congress and the administration.''[28] Fundamental fairness suggested that these wrongs be rectified, particularly after Native Americans' important service in the World War II effort.

Especially needing restitution were the Indian lands that the federal government often had sold without tribal consent. The funds earmarked for this were supposed to be held in trust, but instead they were used to pay for BIA services or were lost or stolen. These indemnities, as they were called, amounted to more than $1.2 billion in legitimate claims. One estimate put 196 tribes living on 200 reservations in 23 states as having lost land and funds through this process.[29]

There were problems for Native Americans when they pressed their claims in federal courts. In 1855, Congress had created the Court of Claims to handle individual and group suits against the United States, primarily because of the doctrine of sovereign immunity, according to which the United States could not be sued without its permission. There were many lawsuits, and Congress tired of special pleading, so it institutionalized the problem. Thus it fell to the Court of Claims to hear Native American complaints about treaty abrogations.

No sooner had the tribes begun to realize that they had a formal outlet for resolving disputes than Congress revised the jurisdiction of the Court of Claims to exclude from its general jurisdiction any claims based on Indian treaties.[30] Instead, Indians had to seek special permission from Congress to bring a suit in the Court of Claims, and Congress first allowed Native Americans to bring specific tribal suits against the federal government in 1881.[31] This action was highly discriminatory because Indian nations rarely had enough funds to lobby Congress. Native Americans had minimal political clout and were unable to pursue lengthy legal battles.[32] By 1950, only 118 separate claims had been filed by Indians in the Court of Claims, and only 29, or 25 percent, resulted in partial restitution for tribes. Most of the Indian cases were dismissed on technicalities.[33] Several cases before the Court of Claims in the 1930s and 1940s proved to be quite controversial. In 1935, because of a veiled Fifth Amendment right of "just compensation," the Creek Nation won compensation for lands unlawfully taken by the United States.[34] Two years later, in *Shoshone Tribe of Indians v. United States* (1937), the court took up a Shoshone claim of an unjust taking of tribal lands, because the United States had settled Arapahoes on the Shoshonis' Wind River Reservation without the Shoshonis' permission. Surprisingly, the Shoshonis won $4.4 million plus interest.[35]

These kinds of victories provided hope for many other tribes. Next to try the Court of Claims were the Seminoles, and their cases established an important legal precedent. In *Seminole Nation v. United States* (1941, 1942), the court ruled that the Seminoles had a strong grievance, but that their damages had to be offset by the money that the federal government had already spent on them. This had a chilling effect on litigation that was transferred to all future claims under the ICC.[36]

By 1946, a certain momentum had gathered in Congress to resolve the Indian claims issue. Morality and New Deal loyalty provided powerful arguments in favor of legislation. State aggressiveness in state courts and the failure of the Court of Claims to handle Indian suits quickly and fairly also contributed. Post–World War II euphoria signaled that it was simply the fair thing to do. After all, it was the age of the Fair Deal.

On August 13, 1946, the Indian Claims Commission Act was signed by President Harry S Truman. The new law contained a variety of important new provisions that profoundly touched Native American rights. And once again, the seeds of destruction were sown in the field of reform. President Truman set the tone at the official signing of the act:

> I hope that this bill will mark the beginning of a new era for our Indian citizens. . . . With the final settlement of all outstanding claims which this measure insures, Indians can take their place without special handicaps or special advantages in the economic life of our nation and share fully in its progress.[37]

What the "special advantages in the economic life of our nation" were Truman did not spell out, but his message was clear. The original purpose of the ICC—that of resolving the legal wrongs committed by the United States against the Indian peoples—somehow became diluted. On the very day of its signing, that lofty goal was enmeshed with the budding termination movement, an extreme acculturationist doctrine that dominated Indian policy and decimated Indian rights in the 1950s.

The Indian Claims Commission Act was composed of twenty-five sections. The commission had three members—the chief commissioner and two associate commissioners—and it had a limited life of no longer than ten years, although the term was extended several times so that the commission officially lasted for more than thirty-two years, from 1946 to 1978. During its first five years, it received all claims. No claims could be filed after the 1951 cutoff. Native Americans were not precluded from relief, but instead they would have to go back before the Court of Claims. The commission had considerable powers: It could adopt any procedures it desired; it could meet when it wanted; and it could authorize "all necessary expenses" to finish the job.[38]

Of particular importance to Native Americans were the jurisdiction, hearing, and appeal sections of the Indian Claims Commission Act. Five kinds of claims were authorized on behalf of Indian tribes or bands, or groups of American Indians. Thus one did not have to be represented by a tribal government to make claims, but if an Indian nation had a tribal government recognized by the Secretary of the Interior, then only that government could bring a claim to the commission. The act, in conjunction with the IRA, brought status and power to tribal governments in one way or the other.[39]

Claims were heard if they arose from treaties or executive orders; if they constituted a tort for which Indians might have sued if the defendant had not been the United States; or if fraud, duress, or mistakes had been a part of treaties, contracts, or agreements between the United States and Indian tribes or bands. In addition, claims from land confiscation without compensation agreed to by the Indian parties and claims "based upon fair and honorable dealings that are not recognized by any existing rule of law or equity" could be adjudicated. Jurisdiction, in essence, was extremely broad. Moreover, no statute of limitations or laches (the inequity of enforcing a claim) could be used by the United States as a defense to dismiss any suit brought by Native Americans.[40]

Congress made sure that this open-ended invitation to sue the United States had some controls, and these provisions proved to be controversial. The commission was authorized to grant relief in terms of monetary awards that Congress would then appropriate. In the process of determining the damages, lands were evaluated by their value at the time of the taking rather than at their 1946 valuation. Offsets were also to be considered, a vestige of the recent *Seminole Nation* decisions. The kinds of offsets allowed included money and property given by the United States to particular tribes and bands. Fortunately, Congress provided for exceptions, such as funds spent on Indian removal, highways to reservations, educational aid, and health benefits. Also, expenditures associated with the IRA or Indian New Deal programs were not counted.[41]

The hearings involved the presentation of claims. Suits already pending in the Court of Claims or the Supreme Court could not be heard by the commission. The offset provisions of the act restricted any cases before those courts. There would be no surprises for Congress such as *Shoshone Tribe of Indians v. United States* entailed. Appeals of commission decisions also went to the Court of Claims, which had the power to reopen each Indian claim.

With the Indian Claims Commission Act, Congress hoped to resolve Native Americans' many simmering disputes. The act attempted to permit a fair hearing of these issues, but there were many problems in the operation of the ICC. Basic questions provoked prolonged debate. The identity of the claimants, the value of the land, the particular grounds for each suit, and the complex nature of specialized evidence placed tremendous strains on the process. The ICC in action lasted for more than thirty years and was nearly institutionalized. But Native American rights and restitution remained uncertain throughout the proceedings.[42]

Assessing the Indian New Deal

Any assessment of the Indian New Deal offers mixed reviews. Such was the experience of the Lower Brulé Sioux, a portion of the Lakota Nation located in central South Dakota on the Missouri River. The BIA saw the Lower Brulé as a test case for the Indian New Deal on the Great Plains, and the agency attempted to apply most of its programs to them.[43]

Originally, the Lower Brulé Reservation contained 446,400 acres, but after allotment, only 234,653 acres were left. The best land was taken by white farmers and ranchers. The land left to the Brulé was better used for cattle ranching, and since the Brulé owned few cattle, most of their tribal lands were leased to whites during the Depression. In 1935, only 7 of 100 Brulé families on the reservation earned a regular income. The overwhelming majority, 93 percent, lived on rations. Thus the Indian New Deal met a very positive reaction among the Brulé.

Under New Deal programs, 35,000 acres were added to the reservation. Seventeen houses were built. Truck gardens with irrigation were introduced, but they failed because of disputes over water rights with non-Indians. No school existed on the Lower Brulé Reservation, because the one that did had been destroyed by a tornado and had not been rebuilt. Children were sent to the neighboring Crow Creek Reservation rather than to boarding schools far away. Health services finally reached the reservation, and this eventually had an impact.

The Lower Brulé approved the IRA in December 1934, when seventy-one voted yes, thirty-nine voted no, and fifty did not vote. A constitution was approved in November 1935, and a charter of incorporation was ratified in July 1936. Under the charter, the Lower Brulé formed the Brulé Cooperative Livestock Association by applying for and receiving a loan from the revolving credit fund. Local white ranchers, especially those with reservation leases, were very upset. In 1939, the BIA restricted the private ownership of cattle among the Lower Brulé, and this led to a number of resignations from the association. The Brulé's livestock cooperative lasted for fourteen years.

Thus some land was returned, and some, although very little, housing was built for the Brulé. Attempts to improve economic development had minimal success. The lack of local cooperation and the failure of the BIA to follow through with the reforms diminished any significant chance at self-sufficiency. Still, some Brulé received training in various programs and benefited from the Indian New Deal, and death rates had dropped by 55 percent by the time of World War II. But overall, after the Indian New Deal was over, there was little to show for it on the Lower Brulé Reservation. The vast majority of Brulé were no better off after World War II than they had been during the Depression.[44]

The Indian New Deal did record a number of accomplishments. The Indian land base was protected and even increased by 4 million acres. Over 7 million acres of leased land was returned to Indian nations, but much of the allotted land remained under the control of non-Indians. Health services helped lower the high death rates; more Indians went to more day schools and received training in postsecondary schools. The BIA hired many more Indians to hold positions on reservations, and some even occupied policymaking positions. The revolving credit fund helped many tribes start economic development projects, especially in ranching. By 1943, John Collier reported that most loans went to full-blood tribal members, thereby reaching those with the greatest needs. Indeed, economic grants had a greater impact on tribal communities than did political reforms.[45]

In terms of Indian rights, however, much ground was lost and some was gained. Greater confusion is perhaps the best assessment. Heinrich Krieger, in a 1935 *George Washington University Law Review* article, summarized this sentiment:

> The Indian is placed under special authority, special protection, special dependency and special independence, not because he is an inhabitant of a certain territory, but because he is a person who, for racial reasons, especially in consequence of the development of Indian law, is in need of the benefits of such special power.[46]

New Deal legislation encouraged the exercise of tribal sovereignty. Tribal governments asserted powers not expressly forbidden by Congress. At the same time, the courts refused to extend the Bill of Rights to Native Americans, and Congress retained certain rights limitations on

Indians that placed Native Americans in a state of continued dependency.

■   ■   ■

It has long been recognized under American law that Indian nations are reservoirs of sovereignty, portions of which are given up by delegation. This came through treaties and agreements and through federal legislation as a result of *Lone Wolf v. Hitchcock* (1903). Among the federal legislation taking powers and rights away from Indians, five laws or administrative rulings stand out: the Resolution of 1871, the Major Crimes Act of 1885, the General Allotment Act of 1887, the decades of forced acculturation administrative pronouncements of the Department of the Interior from the 1880s to the 1920s, and the Indian Reorganization Act of 1934.[47]

During the period of the Indian New Deal, all these laws were the subject of close examination. The Major Crimes Act was expanded, and the General Allotment Act was abolished. Many of the acculturation rulings were reversed, but some remained, and most could be reinstituted because only a few were prohibited through specific legislation. The Indian Reorganization Act was created, and the Resolution of 1871 was narrowed in its scope and implications by the enactment of a law authorizing the formation of the Indian Claims Commission to hear treaty disputes.[48]

The IRA created tribal governments, and it was through them that the rights of Indians could be protected. John Collier favored indirect rule rather than direct rule over these governments. He wanted to use Native American institutions as vehicles for political and social change as long as they could be shaped into democratic communal organizations. If tribes with traditional governments resisted, Collier did not hesitate to crush them. He believed that modern, sterile, duplicative tribal governments were authorized by the IRA and were necessary for positive change.[49] According to New Dealer D'Arcy McNickle, Indian tribal governments had the powers "to function as a municipal body and to exert the common-law rights of a property owner."[50]

Many Indians were realists in regard to reform and the federal government. They were concerned about their precarious legal position once again left exposed by the Indian New Deal. Reactions varied. Clarence Wesley, a Coyotero Apache, remarked:

We feel sorry for some of our Indian friends who still have to put up with [the] authoritarian type of supervision. We would like to hang on to [the] IRA, and to be given a chance to do a good job. It was intended for home rule and that is what we want.[51]

Ramon Roubideaux, a Brulé Sioux attorney, disagreed, observing that home rule was a sham: "Self-government by permission is no self-government at all."[52] D'Arcy McNickle offered probably the most cogent analysis. "The Indian," he explained in a speech delivered before the Missouri Archaeological Society in 1939, "is still at the mercy of his conquerors, and will be for generations to come. But at least he is safe for the time."[53] That time would last for only a decade before termination became a reality.

6

A Cancer Within

The post–World War II fears and frustrations of the United States centered on ferreting out communists wherever they might be found. It was a time of hysteria, and many innocent people were hurt. Strangely enough, "Reds" could be found even on Indian reservations. Although individual Native Americans were not accused of treasonous duplicity, some non-Indians, including national politicians, attacked what they determined to be communistic concepts of Indian law. Amazingly, they saw the ideology of Marxism in Native American culture, and its elimination matched their own private agendas.

There were many cancers from within during the 1950s, and incredibly, many Americans somehow believed that Indian reservations and tribal organizations were manifestations of this disease. The federal doctor ordered termination as a strange kind of cure. Attacks on individual Indians, particularly the young, came in the form of the federal relocation policy. The Indian Claims Commission evolved into an agent of termination, and perhaps the most successful attempt to limit Indian sovereignty and stifle Indian rights came with the passage and implementation of Public Law 280.

By the mid-1960s, the hysteria had diminished, and the realization that communism was not a threat from Indian reservations could safely be argued. The efforts to liquidate Indian lands and sovereignty, the Indian New Deal, and Native American culture had failed. Indians continued to exist, and with the civil rights movement, their many grievances with the majority society became known. The promise of a special bill of rights for Indians was the culmination of a reexamination of two decades of hysteria.

Termination and Triumph:
Surviving Federal Attempts to Liquidate
Native American Culture Through Law, 1947–1967

These twenty years of regression followed an extremely productive era that reversed some of the worst infringements on Native American rights. In some ways, it can be described as a reaction. The Indian New Deal sought to give Indians some control over their own culture while providing tribes with certain legal and political tools to restrain those who opposed this concept. Congress during the 1950s responded with laws that turned young Indians into urban refugees and forced tribal governments to cede judicial and executive powers to the states. The Supreme Court placed Indian nations' sovereignty or very existence in jeopardy, with holdings denying Fifth Amendment rights once again. And the executive branch supported the Indian Claims Commission as it slowly and methodically gutted Native American treaty claims.

All these devastating developments were held together by the glue of the termination movement. It was an Indian nightmare. Although the federal government had the power to enforce it, the question remained whether Native American tribal governments and culture had the will to resist one more time.

"Liquidation"

Liquidation was the initial term coined by the Bureau of Indian Affairs for the ultimate assimilative policy of abolishing Indian tribes altogether, a policy that the BIA championed in the late 1940s. Shortly thereafter, a new word, *termination,* was chosen for the policy's title. Many Native Americans today continue to refer to the policy as liquidation, its original name, much to the BIA's consternation. Although the word "termination" has an ominous finality to it that reminds many people of cancer's last stages, "liquidation" contains even worse connotations that suggest the Nazi horrors of World War II.[1]

Basically defined, liquidation or termination—the commonly used term today—meant the ending of all federal–tribal legal relationships. Native Americans were to be fully assimilated into American life. No legal exceptions were to be made; the trust was over. Federal services were to be stopped. Reservations were to be abolished. Tribal assets no longer existed. Interestingly, it was never clarified in the early debate whether Indians might fall within the penumbra of the Bill of Rights once they were actually terminated.

Termination had its roots in many past policies of the federal government. The goal of the Indian Trade and Intercourse Acts of the late eighteenth and early nineteenth centuries, ostensibly to regulate international commerce and diplomacy, was actually to bring "civilization" to Native Americans. Federal officials implementing Indian removal during the mid-nineteenth century forcibly tried to create a class of Native American refugees. Eventually, land became a key to acculturation, and the General Allotment Act of 1887 emerged as the centerpiece for acculturation designs. It joined other legislation—the Major Crimes Act, the boarding school movement, and the Citizenship Act—to anchor the assault on the fundamental rights of Native Americans.[2]

The harshness and destructiveness that these policies brought to native peoples was recognized in the 1920s, and out of this concern came the Indian Reorganization Act of 1934 and the Indian New Deal programs. However, shortly after World War II, in 1947, the Senate Civil Service Committee held hearings on ways to dismantle the New Deal. A leader on the committee was Senator Arthur Watkins of Utah, who became the prime instigator of the termination movement in Congress.

The Senate Civil Service Committee subpoenaed William Zimmerman, the Acting Commissioner of Indian Affairs who had replaced John Collier, and ordered him to produce a plan for ending the federal–tribal relationship. Zimmerman complied. The Zimmerman Plan divided all Indian nations into three groups: those tribes that supposedly were ready for immediate termination of federal services; those that could handle termination within ten years; and those, after additional governmental expenditures on education, that would eventually be ready for termination within fifty years. The three groups encompassed every federally recognized Native American tribe and band in the United States.[3]

There were three placement criteria. The first involved the degree of acculturation: The BIA assessed the mixture of white blood in individual tribes and bands, the percentage of Indian illiteracy in English, the acceptance of Native Americans by whites in surrounding communities, the willingness of tribes to accept white institutions, and the business acumen of tribes. The second criterion was the BIA's judgment of the ability of tribal members to earn a "reasonably decent living." And third, the BIA recorded the agreeableness of tribes and states to transfers of control. Using these criteria very loosely, the BIA placed 78 tribes and bands in Group I, 20 in Group II, and 78 in Group III. Three bands were left unclassified and probably were placed in Group III. This constituted a total of 179 federally recognized tribes and bands.

Tribes opposed termination as early as 1947. They saw President Harry S Truman disassociating himself from the Indian New Deal, deemphasizing support for tribal governments, embracing termination, and selecting Dillon S. Myer to be the Commissioner of Indian Affairs. Myer's previous experience in government included supervising the relocation camps built to house Japanese-Americans during World War II. Myer strongly favored termination and instructed BIA employees to prepare for it. The trouble for Myer was that Congress was not listening, and so to attract attention, he proposed a 70 percent increase in federal spending for the BIA in 1952. Congress was livid, and in an atmosphere of evil, on August 1, 1953, the House passed Concurrent Resolution 108, the blueprint for termination, without a dissenting vote. The Senate accepted the resolution three days later, and termination became federal policy without debate.[4]

In Resolution 108, Congress ordered wardship to cease. Federal services to Indians also were to stop, with the BIA phased out. Resolution 108 specifically named four states—California, Florida, New York, and Texas—where all tribes were to be terminated. In addition, five specific tribes to be liquidated were named: the Flatheads of Montana, the Klamaths of Oregon, the Menominees of Wisconsin, the Potowatamies of Kansas and Nebraska, and the Chippewas on the Turtle Mountain Reservation in North Dakota. The Secretary of the Interior was asked to submit legislation by January 1, 1954, to carry out the termination of all tribes listed.[5]

Resolution 108, written by Congressman William Henry Harrison of Wyoming and sponsored in the Senate by Henry Jackson of Washington, defies rational understanding. It is not at all clear why four states were singled out or why five separate tribes were designated for termination. The number of tribes specifically mentioned totaled thirteen, although some were not on the Group I list of the Zimmerman Plan. Harrison and Jackson simply provided a pork-barrel political solution.

As Russel L. Barsh and James Youngblood Henderson wrote, Concurrent Resolution 108 and termination involved "a kind of political arrogance to match the cultural arrogance that motivated the allotment policy."[6]

The leader of termination in Congress was Arthur Watkins, chair of the Senate Indian Affairs Subcommittee on Termination. It is difficult to fathom any rational philosophy behind Watkins's views. He believed that Indians had too much and needed to sacrifice and that they should be able to dispose of all their real property if they wanted to, to pay for all physical improvements to their lands, and to be free to experience discrimination. This he termed "full freedom." Writing in 1957, Watkins described Resolution 108 as if it were one of the great documents of the American experience:

> With the aim of "equality before the law" in mind, our course should rightly be no other. Firm and constant consideration for those of Indian ancestry should lead us all to work diligently and carefully for the full realization of their national citizenship with all other Americans. Following in the footsteps of the Emancipation Proclamation of ninety-four years ago, I see the following words emblazoned in letters of fire above the heads of the Indians—THESE PEOPLE SHALL BE FREE![7]

Such are the words of demagogues in Lincoln's clothing.

Termination bills were first filed before Congress in 1954. Each tribe to be terminated had to have a separate bill passed by Congress and signed by the president. The Department of the Interior estimated that it would take two to five years to complete the process.[8] The first termination bill debated by the Senate took place in Watkins's subcommittee and concerned six bands of Paiutes and Shoshonis in Utah located on six small reservations. Although they had not been mentioned in Resolution 108, Watkins wanted them terminated. The Indians tried to oppose the plan, but most did not speak English, and few BIA services reached their reservations. The Triumph Uranium and Oil Company, however, did favor termination. Two bands of Shoshonis were eventually dropped from the bill, but the four Paiute bands were terminated. By the end of 1954, Native Americans in western Oregon, 61 bands and tribes numbering 2,100 Indians; the Menominees in Wisconsin; the Klamaths in Oregon; the Alabamas and Coushattas in Texas; and the Uintah and Ouray Ute Reservation Indians in Utah were liquidated.[9] The last tribe to be terminated was the Northern Poncas of Nebraska in 1962.

Presidents Dwight Eisenhower and John Kennedy presided over the termination movement, signing laws that liquidated over 1.3 million acres belonging to 109 tribes and bands containing approximately 11,500 individuals.[10] From the original Resolution 108, tribes in Florida, New York, and Texas plus the Flatheads in Montana, the Chippewas on the Turtle Mountain Reservation in North Dakota, and the Potowatamies in Kansas and Nebraska avoided termination. Conversely, the western Oregon Indians, Utes, Southern Paiutes, Peorias, Ottawas, Catawbas, Northern Poncas, and Wyandottes—tribes not on Resolution 108 lists—were terminated (Table 1).

Of those Indian nations terminated, the Menominees were by far the

TABLE 1.   Termination of Tribes and Bands, 1954–1966

Group	Number	Acres	State	Termination Statute date	Termination Actual date
Menominee	3,270	33,881	Wisconsin	1954	1961
Klamath	2,133	862,662	Oregon	1954	1961
Western Oregon*	2,081	2,158	Oregon	1954	1956
Alabama–Coushatta	450	3,200	Texas	1954	1955
Mixed Blood Utes	490	211,430	Utah	1954	1961
Southern Paiute	232	42,839	Utah	1954	1957
Lower Lake Rancheria	—	—	California	1956	1956
Peoria	—	—	Oklahoma	1956	1959
Ottawa	630	0	Oklahoma	1956	1959
Wyandotte	934	94	Oklahoma	1956	1959
Coyote Valley Rancheria	—	—	California	1957	1957
California Rancheria Act†	1,107	4,317	California	1958	1961–1970
Catawba	631	834	South Carolina	1959	1962
Northern Ponca	442	834	Nebraska	1962	1966

*Includes 61 tribes and bands. Numbers are aggregates.

†Includes 38 rancherias. Numbers are aggregates.

Source: Charles F. Wilkinson and Eric R. Biggs, "Evolution of the Termination Policy," American Indian Law Review 5, no. 1 (1977): 151.

largest and offered the most resistance. The Menominees were unique in many ways. In the 1950s, they were one of the few tribes that held assets communally and that seemed to be relatively prosperous. They managed a Wisconsin reservation of more than 233,000 acres, with a lumber mill and power plants. Although the BIA provided a school and medical facilities, the Menominees paid for most of these benefits. During the Indian New Deal, the Menominees drew up a constitution and created legislative councils to work with the BIA. There was a great deal of confusion and fighting over these forms of government, and the absence of sanction by the Menominee people left the constitutional forms of representation without a proper foundation for leadership. This played into the terminationists' hands.[11]

In 1951, the Menominees sued the United States and the BIA for forest mismanagement. In the Court of Claims they won a settlement of $7.6 million, but in order to collect, the Menominees needed an act of Congress. The BIA refused to ask Congess to pass a reimbursement act until the Menominees submitted a report on tribal termination. Finally, the Menominees filed a report that was opposed to termination, and the state of Wisconsin also expressed its opposition.

Even though the Bureau of Indian Affairs was upset with the Menominees, the House passed an act authorizing the payment of the settlement, which is where Senator Arthur Watkins entered the scene. Watkins refused to allow the Senate to consider the bill. He went to Wisconsin and visited with the Menominees, informing them that unless they accepted termination, they would not receive their rightful settlement. In the midst of confusion and coercion, 169 Menominees approved the termination request. Hundreds of Menominees did not participate, and those who did thought they were voting on whether to receive a per capita allotment of their settlement package. Watkins quickly pushed the bill through the Senate along with the financial settlement, and the Menominees became the first tribe terminated. For the Menominees, it was termination by spite.[12]

Between 1954 and 1961, the Menominees prepared for liquidation. The end result was the creation of a county, Menominee County. The tribe placed the management of its assets, mainly forests, utilities, and the lumber mill, under the control of a corporation, Menominee Enterprises, Inc., which became the only taxpayer in the county. In order to satisfy state responsibilities, the county forced the corporation to sell assets. The mill deteriorated; the tribal forests were sold to non-Indians for tourism development; and the tribal utilities were taken over by non-Indian companies. Most of the tribe joined the welfare rolls. To survive, the Menominees had to hunt and fish, and because they had been terminated, local officials deemed that they had lost all the protections that their treaties had given them. Accordingly, Menominees were put in jail for violating game laws.

The land sales provoked significant resistance. Several Menominees formed a political action group, the Determination of Rights and Unity for Menominee Stockholders (DRUMS). DRUMS initially protested against the land sales, and it went to court to stop the tribal corporation, no longer controlled by the Menominees, from further asset depletion. Eventually DRUMS took over the corporation board and began to work to restore the federal government's recognition of the Menominees. This finally occurred in 1973. But by the end of this time, the Menominees, once a nearly self-sufficient Indian nation living within the constructs of its peculiar constitutional arrangement with the United States, had had its basic nature challenged. Termination had wreaked terrible havoc on the Menominees.[13]

The nine years of termination, from 1953 to 1962, were a time of great suffering for Native American peoples. Midway through this time, criticism surfaced, but it did not deter those in power from continuing. A war situation also plagued the BIA. Terminationists in place in the agency labeled those tribal governments that tried to resist as "communistic." The Commissioner of Indian Affairs in the Eisenhower administration, Glenn Emmons of New Mexico, commented, "I think on the whole the Indians of this country will someday reach the age of twenty-one. It is probably high time that the Indians begin to plan toward that eventuality."[14]

By 1957, the BIA and Congress were forced to scale down the process. No large tribes were threatened thereafter, and instead the BIA began to plan gradual termination, with the goal of having all Indian nations liquidated by July 4, 1976, just in time for a grand celebration. In what might have constituted a new use for the Bicentennial, termination was eventually restrained, and the goal was not reached.[15]

The effect on the tribes that were terminated was drastic. Because the federal Bill of Rights did not apply to Indian tribes or individuals, it could not be used to protect them from federal termination legislation. Most terminated Indian lands were sold. With the end of the trust relationship, previously needed federal protections from state and non-

Indian individual actions were, in most instances, eliminated. State jurisdiction over the Indians was enforced, and state judicial authority was put in place. Perhaps most important, exemptions from state taxation ended, federal programs for tribes ceased, and federal programs for individual Indians stopped. In essence, tribal sovereignty ended.[16] It is important to note, however, that termination did not abolish the tribe; it simply ended any federal relationship with the tribe. This fundamental principle was established by the Supreme Court in the 1970s.

"Termination," as Seminole historian Donald Fixico wrote, "threatened the very core of American Indian existence—its culture." It was a catastrophic catharsis. George Pierre, tribal chairman of the Colville Confederated Tribes of Washington, called termination "the stroke of doom." It moved, according to Pierre, like "a prairie fire of a pestilence through the Indian Country." It was simply another land grab orchestrated by western political leaders, bankers, ranchers, and timber-company owners. By 1968, 109 tribes and bands had been terminated, and these 11,466 Native Americans had lost 1,362,155 acres of land.[17]

Relocation

The end of tribal recognition by the federal government was not the only form that termination took during the 1950s and 1960s. Congress made other changes as well, and they, too, had important ramifications for all Native Americans, not just those tribes designated for liquidation. One such new federal policy was designed to create refugee populations in off-reservation cities. This particularly destructive policy placed tremendous stress on individual Native Americans and state and local services. It was called *relocation*. The primary proponent of relocation was Commissioner Myer, who believed that Indian schools, tribal governments, and Native American hospitals prevented Indians from achieving "independence." Because termination was proceeding at a rather slow rate, Myer devised an administrative policy to achieve the same goals much more expeditiously while clothing the policy's justification in the guise of providing greater civil rights.

What was relocation? The Bureau of Indian Affairs called it the Voluntary Relocation Program. It was started in 1951, and its main purpose was to redistribute what were termed "surplus" reservation residents to urban areas. In its first relocation effort, the BIA placed Navajos in Denver, Salt Lake City, and Los Angeles. They were given a one-way bus or train ticket, and upon arrival, a BIA relocation worker met them and provided a check for one month's rent, clothing, and groceries; some housing recommendations; and a job. The Navajos were then on their own.[18]

Myer encouraged the BIA to extend this policy to all tribes. Posters and pamphlets were prepared to lure young Indians to the cities. Extravagant and fraudulent promises were made, such as the availability of skilled, lifetime jobs, as well as paid vacations, pension plans, union membership, and health benefits. The reality was much different. Job security was nonexistent. Most of the jobs were for unskilled workers with no training provided. The likely job was as a dishwasher. Housing was substandard, sometimes worse than that on the reservation. Ghettos were created from the concentration of unemployed Native Americans in dilapidated urban dwellings. Alcohol became an overwhelming problem, and unions refused to integrate and allow urban Indians to join. Finally, young Indian children in the immediate post–*Brown v. Board of Education* era were denied education in the cities.[19]

By 1954, the cities designated as relocation sites included also Chicago and Oakland, and all tribes with reservation lands were targeted for relocation. San Francisco, San Jose, St. Louis, Dallas, Cleveland, Oklahoma City, Tulsa, and Minneapolis soon followed as urban relocation centers. Because migration back to the reservation by disappointed Navajos began soon after the initial relocation, the BIA decided to move Indians as far away from their reservations as possible in order to prevent their return. Thousands were processed, and by 1955, 3,000 Southwest Indians alone were living in Chicago.

By 1957, nearly 25,000 Indians had been relocated, and ten years later almost half of all Indians in the United States were living in relocation cities. Relocation cost $403 a person, but the cost in human suffering is impossible to assess. Native Americans earned an average of $66 a week. Some Indians called this a new "extermination program." Youngsters were enticed to the city to die, while the old starved on the reservation. The result was another land grab.

Despite the obstacles to returning, some Native Americans did go back to their reservations, with horror stories. In his Pulitzer Prize–winning novel *House Made of Dawn*, N. Scott Momaday graphically portrays the suffering of individual Indians from southwestern tribes relocated in Los Angeles. Under increasing criticism, the BIA became more devious. It continued to send Indians to faraway cities while refusing to give out names and addresses to relatives who inquired about their loved ones. When Congress asked for information, the BIA refused to collect it.[20]

Central to the relocation's impact was the loss of federal services for Native Americans. Once they left the reservation, they no longer could obtain federal support for education, health, housing, or food. States and cities were now faced with the mammoth problem of providing essential services for a refugee population, and these governmental divisions also were not given additional funds. Indeed, it was these hard-hit governmental constituencies that forced the BIA by the mid-1960s to slow down or suspend various relocation programs.[21]

Nevertheless, the damage was significant, perhaps best symbolized by the fate of Ira Hayes, a Pima who died in 1955 after being relocated to Chicago. Hayes fought in World War II; in fact, he was one of six Marines who placed the American flag on Iwo Jima. After his return to the Pima Reservation in Arizona, he could not find work. He was enticed by local BIA representatives to relocate and was sent to Chicago, where much was made of this Indian hero's having "chosen" to move off the reservation. He was given a job, but he soon turned to drink, despair, and death. In many ways, the destructive power of relocation for Native Americans was represented in its tragic consequences for Ira Hayes.[22]

Relocation was a hardship for many young Indians, and it eventually was stopped. Termination was still implemented, but for only a few tribes. Many tribes succeeded in their delaying tactics. Although both relocation and termination still had a great impact on Indian rights and tribal sovereignty, one could argue that the greatest threat to Indian rights emerged from another McCarthy-era policy foisted on Native Americans. This came in an innocuous statute entitled Public Law 280.

Public Law 280

In August 1953, two weeks after Concurrent Resolution 108 established termination as official federal policy, Public Law 280 was passed and signed into law by President Eisenhower.[23] In many ways, Public Law 280 was symbolic of this era. Vine Deloria, Jr., labeled the 1950s and 1960s as a time when law only complicated and confused the already "ancient schizophrenia of ward–domestic dependent nation ideology" used to place Native Americans within the United States legal system.[24]

Public Law 280 certainly fits this description. It authorized state govern-ments to take criminal and civil jurisdiction away from Indian courts and federal employees officiating on Indian reservations, without Native Americans' consent. It was the most successful legal attack on Indian rights and sovereignty since the adoption of the Constitution.

What specifically did Public Law 280 allow? The states were autho-rized to take jurisdiction over all criminal offenses committed by or against Indians and all civil causes of action in which Indians were parties. Three kinds of states existed under the act: *mandatory states,* six states that had been required to take jurisdiction over Indian country; *disclaimer optional states,* states that had been required to include in their constitutions provisions to limit jurisdiction over Indian reserva-tions and that could amend their constitutions to achieve jurisdiction over Indian country; and *nondisclaimer optional states,* states that needed only to pass a legislative act to assume jurisdiction.[25] Optional states did not have to obtain the Indians' consent or advice before acting to assume jurisdiction.

Public Law 280 provided a context for the depreciation of Indian sovereignty. The states were prohibited from exercising jurisdiction over taxation on real or personal property on Indian reservations or Indian water rights.[26] Furthermore, state jurisdiction could not conflict with federal statutes, treaties, or executive agreements, especially those with respect to hunting and fishing rights. Conversely, Indian tribal laws or customs in conflict with the civil laws of those states assuming juris-diction were voided.[27]

Public Law 280 was almost immediately recognized as bad law. Legalists Russel L. Barsh and James Youngblood Henderson described it as "a monument to congressional ambiguity and indecision."[28] It also has been termed "the final abandonment of the New Deal Indian pro-gram"[29] and "the major piece of assimilationist legislation during this era."[30] Most Indians feared discrimination at the hands of state courts and police, and they worried that treaty protections, favorable Supreme Court holdings, and fishing and hunting rights in particular would be lost. The National Congress of American Indians condemned the legis-lation because it altered the "lives and welfare of Indians without con-sultation and consent of the Indians."[31]

Even President Eisenhower disliked the bill that he signed into law, calling it "un-Christian," and he specifically urged Congress to amend the law to provide for Native Americans' consent. Nevertheless,

Eisenhower did not use his veto power on Public Law 280. Because the western states had pushed hard for it, Eisenhower succumbed. But almost as soon as the mandatory states achieved this victory in the battle over reducing Indian rights, the reality of providing law and order set in. It was expensive. Congress did not provide funds for the states to assume these new duties, and furthermore, it did not give the states the power to take land or subsurface oil and minerals, an omission that was particularly onerous to the western states.[32] By the next session of Congess, efforts had begun to amend Public Law 280.

Despite some grumblings, Public Law 280 was implemented. The mandatory states, of course, had no choice. They were required to assume jurisdiction, and there were problems, the most difficult occurring in Nebraska. Nebraska has three reservations, the two largest—the Omaha and Winnebago reservations—being coterminous along the Missouri River. Nebraska's state government did not want to spend state resources on enforcing Public Law 280, and so it simply ignored it and left the local counties to cope with enforcement. Once federal officials left the reservations, no one was available to enforce law and order. The nearest counties refused to station deputy sheriffs in the main reservation towns of Winnebago on the Winnebago Reservation and Macy on the Omaha Reservation; instead, Indians with problems were required to call the nearest sheriff and hope that someone was there.

Indians on the Omaha and Winnebago reservations soon were at the mercy of criminals. Gustavus White, chairman of the Omahas, reported to state officials that Thurston County officers were not responding. According to White, the reservation was in a state of lawlessness.[33] Complained one Omaha Indian in Macy, "We had some killings going on there, one right on main street, which could have been prevented if we had law and order. This is not exaggerating. . . . How this situation can exist in the United States is beyond me."[34] By 1957, the governor of Nebraska felt compelled to journey to Washington, D.C., to pressure the BIA into taking back jurisdiction, but the agency refused. Finally, the state provided $3,000 in emergency funds to bring limited law enforcement to the reservations. But Nebraska's problems with Public Law 280 were not over.

The optional states also had problems. The process of implementation brought many tribes into the political process of their home states, and they tried to prevent wholesale losses of tribal sovereignty. For example, in 1964, Indians in Wyoming led a successful fight against amending the state constitution. In South Dakota, Native Americans successfully promoted the passage of a referendum revoking the state's legislative attempt to assert jurisdiction over all of South Dakota's reservations.[35] After a number of disputes in Montana, the Montana legislature passed a law whereby tribes that had consented to state jurisdiction could revoke that permission.[36] This procedure, labeled *retrocession,* figured prominently in efforts to modify Public Law 280.

Disputes arose over aggressive state action to assume power beyond the scope of Public Law 280. Eventually, the Supreme Court confronted Public Law 280 in 1959, when deciding *Williams v. Lee.*[37] In this Arizona case, a non-Indian trader who operated the Ganado Trading Post under a federal license sued a Navajo couple in state court for the price of goods sold to them on credit at the store, which was located on the Navajo Reservation. The Arizona Supreme Court ruled that the state could summarily assume criminal and civil jurisdiction over the Navajo Reservation and other reservations in Arizona because Congress had not expressly prohibited Arizona from doing so.

Justice Hugo Black, speaking for the majority, ruled that the Arizona state courts had no jurisdiction in this case and, that the state of Arizona had not followed the procedure for assuming jurisdiction as provided by Public Law 280. Arizona had a disclaimer in its constitution, which it had not amended in order to assume jurisdiction. Black could have stopped there, but he went further, articulating a test for states that hoped to obtain jurisdiction over Indian country: "There can be no doubt that to allow the exercise of state jurisdiction here [Arizona] would undermine the authority of the tribal courts over reservation affairs and hence would infringe on the right of the Indians to govern themselves."[38] Thus states might "take" jurisdiction if they did not unduly interfere with the powers of the tribal governments. Although the Navajos won the *Williams* case, the Navajo Nation was now confronted with the possibility of concurrent state jurisdiction subject to court interpretation, a new loophole that was immediately exploited in Alaska in *Kake v. Egan* (1962)[39] and subsequent cases.

Public Law 280 in many ways was the most successful attempt by the federal government to attack the cancer within. The states could be counted on to try to destroy the nascent forms of Native American tribal governments so recently established by the Indian New Deal. However, those states that had favored allotment did not like Public Law 280. Under allotment, states obtained tribal lands, redistributed them, and decided on those services that they wished to provide. Under Public Law 280 those states that wanted to assume jurisdiction over Indian country quickly found that land could not be cut loose easily, and so they lost funds because they had to provide services without federal support.

By the 1950s, Indians' resistance to the sustained erosion of their rights had reached a low level. Termination, relocation, and Public Law 280 had come full force, but still the worst was yet to come. The Indian Claims Commission prolonged a circuitous, fundamentally unfair process, and the Supreme Court dealt a significant blow to Indian sovereignty and Indian rights.

The Indian Claims Commission in Action, 1947–1967

To some, the purpose of the Indian Claims Commission created in 1946 was to dispense justice for long-overdue tribal grievances. Indeed, it may have been formed in part to "clear the historical slate,"[40] but the ICC was quickly channeled away from providing justice and toward providing an underpinning for tribes on the termination block. Termination, relocation, and the application of Public Law 280—the triple scourge for Native Americans of the Red-scare era—were thought by the federal government and by many of the public concerned with these issues to be programs that could be managed with minimal new state and federal resources, all because of the potential of the ICC's anticipated allocations.[41]

The first commission members were named. President Truman resisted a lobbying effort on behalf of Felix Cohen, who was perhaps the one person who knew the most about Indian claims, and instead nominated Edgar Witt, former lieutenant governor of Texas, as chief commissioner, and Louis J. O'Marr, former attorney general of Wyoming, and William M. Holt, a Nebraska attorney, as ICC commissioners.[42] They were chosen precisely because they lacked experience in Indian law, and the result was a bureaucratic and legal mess.

Before the Indian Claims Commission Act was passed in 1946, the Court of Claims had heard only 142 Indian claims; there was a tremendous backlog of grievances. Thus the ICC was to prepare for the resolution of long-standing claims. The act mandated that claims be filed within five years—that is, by 1951. During that period, 852 separate cases were lodged with the ICC by various tribes, bands, and individual Indians.[43]

The ICC was originally chartered for five years, but it did not complete its job in the allotted time. By the time of the claims cutoff in 1951, the ICC had heard and dismissed only 26 claims and had granted 2 awards. Five years later, only 102 claims had been decided, and Indians had recovered in 21 cases. Approximately $890 million in claims was reduced to $13 million. No fewer than five extensions were needed until the commission finally ended in 1978. Eventually, the number of commissioners was increased from three to five.[44]

After this abysmal start, Congress and Presidents Eisenhower and Kennedy decided that more expertise was needed on the ICC to complete the hearings. Thus when one of the commissioners retired, he was replaced in 1960 by none other than Arthur Watkins, the former senator from Utah and godfather of the termination policy. Watkins left no doubt about the ICC's principal function: He had stated three years earlier that "a basic purpose of Congress in setting up the Indian Claims Commission was to clear the way toward complete freedom of the Indians by assuring a final settlement of all obligations—real or purported—of the federal government to the Indian tribes and other groups."[45] To Watkins, "complete freedom of the Indians" meant termination.

Where did the ICC go wrong? First, although the ICC created an adversarial relationship between the parties, they were not treated equally. Indians had to obtain permission to hire their lawyers; the Department of Justice represented the United States. In an effort to cut down on the Indians' legal costs, their lawyers were restricted to billing no more than 10 percent of the final award. Such a restrictive contingency fee proved not to be enticing to many qualified attorneys. Indians initially had minimal funds, if any, to use to collect data; the Department of Justice had the federal government's largess. The ICC was empowered to set up a Bureau of Investigation to assist in collecting data, but the commission refused to authorize it to function. Thus the Native Americans' side in the proceedings was handicapped from the outset, and in virtually all policy issues, the ICC came down strongly on the side of the federal government.

In the first decision rendered by the ICC, the Fort Sill Apaches were denied their claim of unfair imprisonment. They alleged that they had been placed in jail solely because of the actions of Geronimo and several others. But the commission ruled that the Fort Sill Apaches represented an aggregate of individuals rather than an identifiable Indian group,[46] a decision that had a chilling effect on subsequent litigation. The definition of a representative Indian group was narrowly construed and required significant, perhaps overwhelming documentation in the face of complex, entangled tribal histories. Many tribes and bands were forced to live on reservations with others and were prohibited from observing traditional cultural practices. Only those, therefore, that had successfully resisted the forced acculturationists could prove continuous identity.

Other complex issues were the establishment of title to lands and the value of the lands. The commission ruled that Native American claims could not overlap geographically, but by definition, this rule was absurd. Homelands were frequently overlapping, and buffer zones of mutual interest were common among Indian nations. Once a fraudulent taking was proved and the title to the land was established, the value of the claim had to be assessed. The commission, with congressional approval, made decisions based on the notion that the value of dispossessed lands should be predicated on the lands' worth at the time of the taking. The commission then became embroiled in controversies over what dates to use and what real-estate surveys to adopt. On more than one occasion, the Department of Justice succeeded in establishing inaccurate dates and deflated land-value assessments. Those few decisions appealed to the Court of Claims were sent back to the ICC with admonitions to the commissioners not to accept such shoddy evidence.[47]

Two important Bill of Rights legal concepts were trampled by the ICC: due process *res judicata* restrictions, and takings without just compensation provisions in the Fifth Amendment. When the ICC first began to hear cases, the earlier settlement of a claim by a tribe was not to preclude a new day in court. Previous holdings in the Court of Claims were based on narrow grounds. Under the Indian Claims Commission Act, much broader claims could be heard, but this provision was jeopardized when the commission ruled in *Choctaw Nation v. United States* (1950) that a new claim brought by the Choctaws had already been considered and so could not be reopened.[48] Thereafter, a number of claims were dismissed.

Similarly, the Fifth Amendment requires just compensation after a taking by the federal government. The ICC was asked to rule whether interest for land taken accrued from the date of the taking. Having pushed the evaluation process back into the nineteenth century, the commission now faced the enhancement of the awards by computed compound interest. The ICC refused to approve the payment of interest, and finally the Supreme Court ruled that unless Congress specifically assumed liability, it did not have to pay it. The ICC was therefore upheld in its desire to lower the awards and restrict the application of the Fifth Amendment to Indian nations.[49]

Throughout this process, the Department of Justice acted reprehensibly. Under the guise of an adversarial system, lawyers for the federal government resorted to outrageous tactics. Delay after delay was used to try to force Indians to settle out of court. For example, it took the Yankton Sioux twenty-four years, from 1951 to 1975, to obtain two checks, for $249 and $930.86 for each tribal member. The concept of offsets was stretched beyond rational recognition. The ICC allowed the federal government to mitigate the costs of preliminary awards. Called *gratuitous offsets,* this accounting took into consideration those expenditures previously paid to Indian tribes. For example, the attorney general's office argued and won from the ICC an offset for the funeral expenses of indigent Indians, from a successful claim by the Quapaws. Arguments were developed to offset the cost of food given to starving Indians after their initial placement on reservations.[50] Thomas LeDuc, a historian, summarized the role of the Department of Justice in ICC proceedings: "One could say that official chicanery at the expense of the Indians has merely been transferred from the Indian Bureau and the Army to the Justice Department."[51]

By 1968, the Indian Claims Commission represented a miserable failure for Native Americans. The process was self-defeating. Indians who attempted to litigate found themselves fearful of termination. Tribes preferred to distribute any funds rather than submit plans that could be used against them in a termination battle, and by the time the federal government finished with the offsets, little was left to distribute. Deciding on the use of tribal funds provoked divisive battles that threatened the very foundations of tribal governments on reservations. By 1968, the ICC had proved to be a strong new cancer from within.

There were some positive signs, but they were few. After a long, hard-fought battle, the Kiowa, Comanche, and Apache tribes succeeded in persuading the ICC that depredations, the loss of life by settlers during Indian–United States confrontations, could not be counted as gratuitous offsets, nor could payments for food and other supplies to indigent Indian survivors of massacres and United States Army raids.[52] This was a major breakthrough, but it came too late for many tribes. The

ICC process also provided a political by-product for native peoples. Political leadership in the tribes developed around finding techniques for working with the federal government and for administering tribal funds. Considerable expertise was required, and those tribes that won awards did so only with great persistence.

It should not be forgotten, however, that the ICC did not finish the process of resolving Indian claims. Fundamentally, the tribes and the United States became adversaries, and reasonable compromises were impossible to achieve. The ICC process eventually simply exhausted Congress by the time it came to a close in 1978.[53] Not surprisingly, the Bill of Rights was not invoked to help Native Americans survive or even cushion the impact of the ICC in action.

Tee-Hit-Ton Indians v. United States *(1955)*

In the midst of the operation of the Indian Claims Commission, the Supreme Court issued a new pronouncement regarding Indians and their relationship to one of the articles in the Bill of Rights, the Fifth Amendment. This decision profoundly affected not only ICC proceedings, but Indian law in general.

The dispute involved the Tee-Hit-Ton Indians of Alaska, a small separate confederated division of the Tlinget Tribe, and its claims that the Department of Agriculture cut, without compensation, timber from the Tee-Hit-Ton traditional homelands.[54] At the time of the taking, the Tee-Hit-Ton numbered approximately seventy members, and their homelands stretched over 350,000 acres of land and 150 square miles of water. The Tee-Hit-Ton did not have a specific treaty agreement with the United States, but they had continuously occupied and used the lands in question before the Russian occupation of Alaska and its purchase by the United States. In addition, the Tee-Hit-Ton argued that the 1884 organic act creating Alaska Territory and the 1900 act establishing civil government in Alaska recognized Indian ownership of lands in Alaska. This allowed them to claim full proprietary ownership. Thus if the Tee-Hit-Ton owned the land, the Department of Agriculture's action constituted a taking without just compensation, a violation of the Fifth Amendment to the Constitution.

The Tee-Hit-Ton brought their claim to the Court of Claims, which agreed that they were an identifiable Indian group that had standing. This part of the holding corresponded to the provisions for those tribes and groups using the Indian Claims Commission Act. But this was all that the court would grant with reference to the Tee-Hit-Ton's claim. The Tee-Hit-Ton then appealed to the Supreme Court.

In 1954, the Supreme Court heard arguments and the next year decided *Tee-Hit-Ton v. United States*. The historical record shows that the Court was very concerned about this case and the Indian Claims Commission proceedings. In the majority opinion, the Court refers to the ICC twice, and it cites as precedent several opinions written by the commission. Reducing tribal claims was to this Court an important point of law that would have an impact on the ICC.

Justice Stanley Reed issued the majority opinion for six members of the Court. First, he summarized the argument by Alaska's Tlingits: The Tee-Hit-Ton believed that they had full proprietary ownership of the timber lands in question near the Tongass National Forest. If the Court was unwilling to accept this interpretation, the Tee-Hit-Ton also contended that they held unrestricted possession, occupation, and use of the lands. The first argument placed the title of the lands in the hands of the Tee-Hit-Ton, almost like adverse possession of property under English common law. The second position accorded reservationlike status to the lands and amounted to the constructive creation of a reservation for the Tee-Hit-Ton.[55]

Reed, however, refused to accept either Indian argument and instead adopted the federal government's reasoning. The United States, backed by amicus briefs from the attorneys general of Idaho, New Mexico, and Utah, reasoned that Indians without explicit treaties or executive agreements had only the right to use lands subject to Congress's will. Reed set up a test: To prove title to lands, Native Americans had to show "definite intention by congressional action or authority to accord legal rights." In short, they had no legal interest in lands unless recognized by Congress, and therefore no compensation could follow.[56]

The majority then went further, concluding that Indians had exercised sovereignty over lands in North America before the United States became a nation. But after conquest by the United States, Native Americans retained only a right of occupancy. This, Reed pontificated, had constituted the established policy of the United States and its courts since the origins of the Constitution. Moreover, he quoted Justice Stephen J. Field, a nineteenth-century judge known for expressing his hostilities toward nonwhites in his opinions. Reed referred to *Beecher v. Wetherby* (1877), a case in which the Court allowed the United States to give Indian lands to the state of Wisconsin. "It is to be presumed," wrote Field, "that in this matter the United States would be governed by such considerations of justice as would control a Christian people in their treatment of an ignorant and dependent race."[57]

Reed concluded that "no case in this Court has ever held that taking of Indian title or use by Congress required compensation." This flew in the face of *United States v. Alcea Band of Tillamooks* (1951), in which the Court only four years earlier had stated that unrecognized Indian title could be compensated if lands were taken by the United States. The Court later spelled this out by adding that Congress had to make the payments and that this was a proper political issue for Congress, not the Court. In effect, Reed overruled the limited *Tillamooks* precedent.[58] The Tee-Hit-Ton therefore lost. Reed hinted that Congress might want to take up the matter, and eventually Congress did in the Alaska Native Claims Settlement Act of 1971.

The significance of the *Tee-Hit-Ton* case is profound, as it came at a time when termination was in its childhood. The language of the Court and the impact of the decision sent a message to all tribes and to the ICC: Native Americans with treaties held certain privileges that may or may not be retained over time; tribes without treaties simply had no legal footing. In effect, all tribes, even those with treaties, were subject to termination.[59]

The ICC and the United States Court of Claims quickly came to the defense of the ICC process, although in a limited fashion. That same year, in *Otoe and Missouria Tribe of Indians v. United States* (1955), the Court of Claims held that Indians could pursue in the ICC claims based on "Indian" or aboriginal title, like that of the Tee-Hit-Ton. However, the *Otoe and Missouria* case and subsequent decisions narrowly defined what this meant: Aboriginal title to lands required reasonable exclusive use and occupancy of lands for time immemorial (or a long time). The lands had to have been occupied before the United States entered the lands, and the tribe could not voluntarily leave the lands simply to escape a genocidal neighbor. Thus if a tribe could meet these four criteria, then a taking might be considered, and compensation was an available remedy.[60]

Tee-Hit-Ton v. United States left a powerful legacy in Indian law and the relationship of Native Americans to the Bill of Rights. Clearly, the Fifth Amendment did not apply to Indian claims of real property losses, which meant that takings could be sanctified by the Supreme Court without compensation. Only the ICC could make available to Indians any limited compensation, and this was regarded as preliminary to termination. Thus the *Tee-Hit-Ton* holding and the workings of the Indian Claims Commission eliminated the Fifth Amendment as a possible right available to Native Americans.

The New Trail

Miraculously, the times were achanging (somewhat) for Native Americans during the 1960s. These changes were in part brought about by the Kennedy administration in the early years of the decade. The application of the Bill of Rights continued to be denied to Native Americans for most important matters of legal concern, but there clearly was a new interest in Congress in finding ways in which Indians might be included in the general civil rights revolution.

The Supreme Court became quite active in issuing opinions on a wide variety of Native American disputes, largely because of a maturing of tribal governments and a realization by tribes and individual Indians that the court system was more open to them than it had been before. Expertise derived from the Indian Claims Commission hearings was put to new use. The Court wrote significant decisions concerning religious freedoms, taxation, water rights, and due process.

Much of the legal controversy over religion for Indians centered on the Native American Church, which combines Christian ideology with traditional Native American religious practices, including the use of peyote. Peyote is derived from a cactus found mainly in the desert areas of Texas and northern Mexico. The buttons of the cactus are ingested during Native American Church ceremonies, and they cause purgation and hallucinogenic visions. Peyote had been used as a religious sacrament among Indians in North America for centuries, and its combination with Christianity by Indians in the United States during the late nineteenth and early twentieth centuries reflects the cultural transitions and adaptations made by many Native Americans.[61]

Not everyone appreciated the cultural implications of the Native American Church or the right of its members to the free exercise of their religion. State officials, particularly in the Southwest, and some tribal government leaders made war on the Church. Simply stated, the First Amendment provisions applying to religion protect religious beliefs. Although they allow for religious practices to be regulated, it is not always easy to determine what these fundamental beliefs or practices are.[62]

In the late 1950s, the Navajo Tribal Council did not want the Native American Church to function on the Navajo Reservation, and so it adopted an ordinance forbidding the Native American Church to hold services and punishing persons possessing peyote with imprisonment and fines. The Church responded by seeking an injunction against the tribal council to prevent enforcement of the tribal law. The dispute went as far as the Tenth Circuit Court of Appeals. There, in *Native American Church v. Navajo Tribal Council* (1959), the court ruled that the tribal ordinance was constitutional because the First Amendment did not apply to the Navajo Tribal Council, and so it could not be used to protect individual Native Americans from having their religious free-exercise rights violated. The court found that tribal governments, although created by federal statutes, were not covered by the First Amendment; rather, it applied strictly to congressional action. Ruling out any marriage of the Fourteenth Amendment and the First Amendment with reference to tribal governments, the court stated that tribes "have a status higher than that of states."[63]

Emboldened by the *Navajo Tribal Council* decision, the states began to attack Native American Church practices. In Needles, California, several Navajos met in a hogan to perform the sacrament of the Native American Church. Police officers present arrested the Navajos and charged them with violating a California statute prohibiting the possession of peyote. The California Supreme Court ruled in *People v. Woody* (1964) that the First through the Fourteenth Amendments did apply to state action. Here the state law regulated religious practices, and the court found that for the law to be upheld, it had to meet a test of a "compelling state interest."[64] All parties agreed that peyote was a substance not physically harmful to those taking it, although the state lamely argued that some Navajos were using peyote instead of seeking medical help. Thus the California court ruled that no compelling state interest justified restricting this particular free exercise of religion by Native Americans.

Issues of taxation came to the fore during the 1950s and 1960s. Working out the Indians' place in federal and state taxation schemes became complex. Before 1946, the federal government could clearly tax Indian income; both Indians and non-Indians living on reservations paid federal income taxes. There were, however, certain kinds of taxes that Congress exempted Native Americans from paying. For example, in 1956 when the federal government tried to assess a capital gains tax on timber it sold from allotted lands on the Quinault Indian Reservation in the state of Washington, the Supreme Court held that such income was exempt because of the General Allotment Act of 1887.[65] But ten years later, the income of Indians who leased tribal trust land was held to be taxable for federal income tax purposes.[66]

The states also wanted to tax reservations and their occupants, although there are significant limitations on the states' taxation powers. In the nineteenth century, states tried to assess property taxes on Indians, but they were unsuccessful: Congress and the courts prevented the states from taxing Indian trust lands. But what about income, both Indian and non-Indian, on reservations? This issue came to a head in the Southwest when Arizona enacted a 2 percent tax on the sales of traders on Indian reservations. The Warren Trading Post, owned by a non-Indian on the Navajo Reservation, challenged the statute, and the Supreme Court ruled that Arizona's law was unconstitutional because federal statutes and policies, such as trader licensing, preempted the states' powers to tax.[67] Normally, the states can tax the income and property of non-Indians on reservations, but they cannot tax Indians or non-Indians when such a tax would infringe on the tribal governments' rights to make laws. Clearly, by the end of the 1960s, confusion reigned over federal and state taxation of Indians and non-Indians on reservations.[68]

Water rights for Native Americans were better defined. Earlier, in 1908, the Supreme Court had held that the creation of reservations implied that Indians had claims to any water adjoining the reservations. Known as the *Winters* Doctrine, this significant legal policy was held in abeyance until the 1950s when Arizona and California fought over the quantity of water to be taken from the Colorado River and its tributaries.[69] The dispute eventually spread to three other states—Nevada, New Mexico, and Utah—as well as five Indian reservations.

In *Arizona v. California* (1963), the Supreme Court upheld the constitutionality of the *Winters* Doctrine and further elaborated on its application. The water rights of Native Americans on reservations stood against all other rights from the time of the reservations' creation. In addition, the quantity of water reserved was based not on the small number of Indians living or likely to be living on a reservation, but on the amount needed to irrigate the tillable acreage on the entire reservation.[70] This constituted a considerably greater amount of water reserved for Indians.

Thus Indian water rights acquired both appropriation and riparian aspects. What was important to Native Americans was that the lifeblood of their survival on reservations, especially those in the West, had been preserved for the time being.

The Supreme Court also extended criminal rights to Indians, primarily in the area of due process and federal habeas corpus applications. On the Fort Belknap Reservation in Montana, Madeline Colliflower, a Gros Ventre, was arrested for having violated an Indian court order to keep her cattle off portions of the reservation. She was put in jail and then sought a habeas corpus writ because she alleged that her confinement was illegal, since she had been denied counsel, a trial, and the right to confront witnesses against her in court. Applications to Indians of the Fifth and Sixth Amendments were at stake.

In a peculiar opinion, the Ninth Circuit Court of Appeals agreed that Colliflower was entitled to a writ. The majority opinion took great pains to state that this was a special one-time-only situation, that had the events happened years earlier or later the court might not have acted, and that the nature of the court at Fort Belknap allowed intervention. The court did not consider the merits of Colliflower's contention. Thus the tribal courts might be federal in some kinds of relationships, but it was unclear as to what kinds. No blanket Bill of Rights application of criminal due process was attached to tribal government action.[71]

The Kennedy–Johnson administration of the 1960s set about to redefine Indian relationships within the federal legal system. What evolved was a policy preference for Indian self-determination. Kennedy initially proposed a ''New Trail'' for Native Americans, and he appointed Stewart L. Udall, a congressman from Arizona, as the Secretary of the Interior. Kennedy and Udall promised to end termination, but that did not stop either from implementing the Northern Ponca termination in 1962. As late as 1967, Udall was appearing before Congress and extolling economic programs to enable abolishing the reservations. If anything, the Kennedy/Udall–led federal government, while professing opposition, chose to go slowly rather than abolish termination and other destructive programs.

What did become a keystone of executive policy during the 1960s was increased economic aid to Indians. By 1968, approximately $35 million had been spent on federal poverty programs for Indians. Although impressive, this constituted only 8 percent of all federal spending in the Office of Economic Opportunity. Conditions on reservations remained deplorable. Unemployment during the 1960s remained at 40 percent. Depending on the season, 95 percent of the Pine Ridge Indians were unemployed. In 1968, the average family income of Native Americans was $500 a year, whereas the average white family earned $5,893. High-school dropout rates were at 60 percent, an improvement from the past but still above acceptable levels. There were 63,000 Indians living on reservations without running water.[72]

Indians were not successful in helping the civil rights movement find a special place for them, and they were unable to persuade Congress to abolish termination or relocation. They could not obtain favorable, fair decisions from the Indian Claims Commission. And although they were included in federal antipoverty programs, they were not a centerpiece or a primary beneficiary of them.

Given the conditions prevailing on reservations and the rhetoric without results, many Indians turned to nationalistic solutions to problems. In April 1959, representatives of the Seminole, Onondaga, Chippewa, Mohawk, Seneca, and Tuscarora nations met in Florida to sign a plan to form a United Indian Nation. Election disputes and civil rights complaints had polarized the Navajo and Rosebud reservations. In 1960, the Hopis appealed to the United Nations for help against the United States because of oil and coal development being forced on them. The Utes declared secession from the United States over a land dispute, and the Chippewas also testified before the United Nations about a Department of Defense plan to place bombing ranges on their reservation lands.[73] Indeed, nationalistic movements were growing rapidly among American Indians. The cancer identified in 1946 had transformed itself into something quite different by 1968.

■   ■   ■

Linda Hogan, in her novel *Mean Spirit,* a story about the slow destruction of the Osage people, recalls their despair after the sudden, mysterious death of one of their children:

> Lying in bed that night in Oklahoma, [Lionel] Tall still grieved. He remembered the body of a small girl whose cap had been embroidered and beaded with the American flag. She lay there, one of her blue hands stretched out, as if asking for help. Uncle Sam was a cold uncle with a mean soul and a cruel spirit.[74]

During the crucial decades following World War II, Uncle Sam displayed an uncommon legal meanness and cruelty toward the native peoples of the United States. This era represented the greatest challenge yet to Native American rights. Termination, relocation, Public Law 280, the Indian Claims Commission, and other federal programs were implemented. In some ways, they were a guise—lacking the candor of the nineteenth-century forced acculturationists—perpetrated to abolish Indians as Indians and to keep them from realizing their full rights as United States residents and citizens.

In this process, there were unlikely non-Indian villains and heroes. Disaster for Native Americans was encompassed in the presidencies of Harry Truman, Dwight Eisenhower, and John Kennedy. Only Lyndon Johnson near the end of his term advocated the abolition of termination. No one wanted to stop relocation, although the Kennedy administration renamed it the Employment Assistance Program, so it, too, sounded better. In fact, both American political parties embraced it. Public Law 280 was accepted by all administrations, and the ICC continued to function, meet, and be renewed. Hank Adams, an Assiniboine writer and activist, observed: "When you realize what the good guys have done to Indian people, then you cannot accept the way things are now and how things are moving toward the future."[75]

The year 1968 brought significant social and political changes to America and profound legal change for Native Americans. This year highlighted the legislative leadership of a little-known segregationist senator from North Carolina, and it also brought into the executive a time-worn politician from California. Who could have predicted that these two would have such a significant legal impact on Native Americans? By December 1968, a different kind of bill of rights was law, and new concepts in Indian self-determination were close to becoming national policy.

7

The Indian Bill of Rights

The social and political upheaval in the United States peaked in 1968. The year began with a presidential announcement of the drafting of over 300,000 men for the Vietnam War, including many Native Americans. The war itself was not going well; the Viet Cong besieged the United States Marine base at Khe Sanh throughout February. One month later, Lyndon Johnson announced to a stunned nation that he would not run for another term as president. Earlier he had fared poorly in the New Hampshire presidential primary against Senator Eugene McCarthy of Minnesota, a critic of the Vietnam War. The violence abroad came home with the assassination of Martin Luther King, Jr., on April 4 and the resulting urban riots. One week later, Congress responded with the passage of another civil rights act, this one aimed at preventing discrimination in the sale and rental of housing.

Students all over the country tried to take over their colleges, and they marched in the streets demanding an end to the war and to discrimination. Before he was assassinated, Martin Luther King, Jr., had planned the Poor People's March in Washington, D.C., to dramatize the extensive poverty in the country, and in May the Poor People's Campaign, led by the Reverend Ralph Abernathy and composed of many poor people, mostly African Americans, converged on Washington and camped near the Washington Monument. On June 5, Robert Kennedy, senator from New York and former attorney general, won the important California Democratic primary, edging out Eugene McCarthy, but that evening he was fatally shot after making a victory speech.

During the late summer, the main political parties held their conventions. In early August, the Republican party nominated Richard Nixon, Eisenhower's vice president and the 1960 presidential nominee, as its standard bearer. During the last week in August, the Democrats gathered in Chicago to choose their candidate. After Johnson's withdrawal, his vice president, Hubert Humphrey, declared his candidacy, and their supporters controlled the Chicago convention, nominating Humphrey on the first ballot. Although Humphrey prevailed at the convention, he and his supporter, Chicago mayor Richard J. Daley, could not prevent bloodshed and mayhem on the city's streets. Hundreds of disappointed anti–Vietnam War protesters were attacked by Chicago police, and the bedlam came into the hall during a famous speech to the delegates by Senator Abraham Ribicoff of Connecticut. In October, the vicious third-party candidacy of segregationist George Wallace of Alabama emerged, but on election day Richard Nixon eked out a narrow presidential victory. By the end of 1968, the nation was trying to come to terms with its horrors, including the deaths of 30,000 Americans in Southeast Asia, by celebrating the success of *Apollo 8* in its orbiting of the moon. It had truly been a year to remember.

Native Americans also remember 1968, and those memories are both complex and vivid. The 1968 Congress, federal and state judiciaries, President Johnson, candidate Nixon, and the Civil Rights Act of 1968 were important to Indians as well. Moreover, also occurring in 1968 was the surprise creation and passage into law of the Indian Bill of Rights, a statute that significantly changed the legal relationships of Native Americans. Issues not yet resolved—from the Indian Claims Commission and termination to relocation and the applications of Public Law 280—were also addressed in 1968. In the end, a new federal policy—that of Indian self-determination—founded on old ideas, was introduced, encompassing a different kind of hope for a sustained legal parity.

Mandate and Morass:
Setting a New Agenda from Old Legalisms,
1968

American history is not filled with numerous periods of upheaval. Although social change has occurred in the United States, there have not been the kinds of violent movements in the society as there have been in others, such as in Russia, China, the Middle East, or Ireland. So when things fell apart in the United States in 1968, the American people and scholars looked for reasons to explain this unusual time.

The year 1968 has been and will continue to be the subject of numerous histories and films. It is featured in texts as a time of troubles when the fabric of American society began to unravel. But within this societal dysfunction was an era of reform. The anti–Vietnam War movement acted as an umbrella over numerous calls for social change, including women's rights and civil rights. Native Americans were also involved in the rhetoric and action of reform, but they did not speak loudly or with unanimity. Moreover, the larger national movement did not address the legal disparities of Native Americans. Indian issues were not issues of the civil rights movements. Instead, women, African Americans, and Latinos came to the fore with their agendas for social change.

Thus for Native Americans, 1968 was an ambivalent time. It was not so much the violence, the war, and the assassinations that made Indians cautious; some remember vividly the Sand Creek Massacre, the numerous wars with the United States Army, and the political murders of Crazy Horse and Sitting Bull. Here in 1968 was a movement in history during which the racial dimension to American society was carefully scrutinized, and Indians were being left out. Then, much to everyone's surprise, an Indian Bill of Rights was born.

Legal Highlights for Native Americans in 1968

Shortly before he was murdered in 1877, Crazy Horse of the Oglalas reportedly observed the following about American power and the defeat of his people by the United States government: "We preferred our own way of living. . . . All we wanted was peace and to be left alone."[1] Nearly a century later, these words might easily be applied to the extension of American law over Indians. In 1968, Congress and other political and legal institutions were not about to leave Native Americans alone.

During 1968, the federal and state governments were especially active in assessing Indian law. The first four months saw Congress pass nine pieces of Indian legislation, the first eight concerning specific tribes and a variety of funding and land issues. The ninth piece of Indian legislation was a last-minute amendment to the Civil Rights Act of 1968. This was the Indian Bill of Rights, a pet project of Senator Sam Ervin of North Carolina. Since 1961, Ervin had held hearings on the legal rights of

Native Americans, and he had become convinced that Indians should have the same individual rights as those guaranteed to all other Americans by the Bill of Rights.

Assessing his motivation is not easy. Clearly, Ervin listened to Helen Scheirbeck, a Lumbee from Ervin's home state who was his aide on the Subcommittee on Constitutional Rights. Ervin admired the acculturation of the nonreservation Lumbees and saw them as models for other Indians. But Ervin was also a segregationist who wanted to delay or stop the extension of civil rights to African Americans. By trumpeting the lack of individual Indian rights, Ervin might have thought that he could build a coalition with western senators to prevent general civil rights legislation from passing.[2]

To be fair, however, we must add that Ervin felt strongly that all Americans deserved the protection of the Bill of Rights. He later stated that "even though the Indians are the first Americans, the national policy relating to them has been shamefully different from that relating to other minorities."[3] The problem was that most Indians believed they were different and wanted to stay that way. Ervin and other senators were quite surprised to learn that the Bill of Rights in 1968 did not apply to Indians living on reservations. Such a notion was "alien to popular concepts of American jurisprudence."[4]

Presidential leadership sought to define Indian programs and legal relationships during 1968. In March, President Johnson, seeking to distance himself and his administration from the Kennedy and Eisenhower years, sent a message to Congress declaring that Indian policy should be guided by a doctrine of self-determination, not termination.[5] Johnson then issued an executive order establishing the Council on Indian Opportunity, whose primary mission was to coordinate federal programs and create a full partnership between Indians and the federal government.[6] During the presidential campaign, Richard Nixon echoed these sentiments. He opposed termination, favored self-determination, and supported the Council on Indian Opportunity.[7]

Like the other branches of government, the judiciary—including the Supreme Court, other federal courts, and the state courts—was also active on Indian issues. The effect of termination on the overall sovereign rights of the Menominees was decided.[8] Clarifications regarding the extent of rulings by the Indian Claims Commission were made.[9] Public Law 280, the power to tax on reservations, and extradition rights were the subject of litigation.[10] Limitations on hunting and fishing treaty rights were beginning to be heard by federal courts, as Indians, particularly in the Pacific Northwest, sought to test state regulations restricting long-standing hunting and fishing practices.[11] By the end of 1968, even the new Indian Bill of Rights was the subject of a dispute over its interpretation.[12] This was a busy time.

Lingering Legal Entanglements

Fresh from another renewal of its existence, the Indian Claims Commission continued to review allegations by Native Americans about the abridgment of treaty rights and human rights. The commission's procedures still hamstrung the process. In an effort to move ahead more quickly on the huge backlog, starting in 1968 the commission allowed one commissioner, rather than three, to hear a case. In addition, pretrial activities were tightened, and experts were required for the first time to submit written reports before a trial.[13] Despite these improvements, the commission's decisions and those appealed to the Court of Claims provoked controversy.

The Peoria Tribe in Oklahoma made a treaty with the United States in 1854, in which the Peorias ceded tribal lands to the United States on the condition that the federal government would sell them at a public auction with the proceeds invested in bonds, whose interest would be annually paid to the tribe. The United States violated the treaty. It sold the lands to private individuals, not at a public auction, and it did not invest the proceeds. When the Indian Claims Commission was established, the Peorias made their claim. At a hearing, the ICC held that the Peorias were entitled to only the difference between what the lands would have sold for at a public auction in 1854 and the funds actually received from the private sales. This amounted to $172,726. The Peorias appealed to the Court of Claims, which affirmed the ICC's decision, and the Peorias then appealed to the Supreme Court.

The federal government argued before the Supreme Court that it was established law for Indian claims that the United States was not liable for interest on claims against it. In *Peoria Tribe of Indians of Oklahoma et al. v. United States* (1968), the Supreme Court, however, refused to accept this argument, instead finding that there were specific treaty obligations to invest the funds and pay an annual income to the Peorias. In a brief unanimous opinion by Justice Potter Stewart, the Court remanded the cases back to the ICC to determine the amount of damages to be awarded to the Peorias. The ICC and the Court of Claims were thus curtly reversed.[14]

The Justice Department seemed intent on tightening any potential loopholes in ICC proceedings. Two small tribes in Arizona, the Gila River Pimas and the Maricopas, attempted to sue the United States for inadequate educational, health, and administrative services provided to them through the years based on the jurisdiction in the Indian Claims Commission Act requiring ''fair and honorable dealings'' by the federal government in its relationships with Indians. Here the Justice Department vociferously argued that the United States had no specific responsibility for providing incompetent services. The ICC ruled that no compensation could be paid because there was no specific standard of care for providing these federal services. The commission ignored the Supreme Court holdings in *Sweatt v. Painter* (1950) and *Brown v. Board of Education* (1954). African Americans were entitled to equal educational opportunities, but Indians were not. The Gila River Pima–Maricopa community appealed this decision to the Court of Claims, which supported the ICC.[15] The commission clearly did not want to open the door for tribes to claim that they had historically experienced damages as a result of incompetent teachers and doctors being sent to the reservations.

The ICC also heard the case of the Northern Paiute Nation, which claimed that their lands in Nevada, Idaho, Oregon, and California had been taken from them by the United States without proper compensation. The commission agreed with the Paiutes' claims and assessed more than $15.7 million in damages. What made the claims so valuable was the finding that the Comstock Lode in Nevada was located on the Paiutes' homelands.

The federal government appealed the decision to the Court of Claims, and the Paiutes filed a counterclaim. The court reached its holding in April 1968, dismissing both claims and reprimanding the Justice Department for trying to argue that the Treaty of Guadalupe Hidalgo extinguished all Indian claims to subsurface minerals. The court went into great detail regarding Spanish mineral law and showed where the government was deliberately misleading in its presentation of the evidence. On the other side, the Court of Claims had little patience for the Paiute attorneys, who tried to assess the same evaluation standards applied to the Comstock Lode for all the silver mines in the region. This made little sense to the court and smacked of lawyerly greed. Thus Indian mineral rights were not eliminated, nor were irrational evaluations for worthless mines allowed as compensation.[16]

In 1968, the Indian Claims Commission continued to process claims and make rulings on narrow grounds. Its findings appeared to be modestly favorable to Native Americans when compared with the aggressive challenges made by the federal government's lawyers. Still the Supreme Court, such as in the *Peoria* decision, had to watch the commission carefully to ensure some semblance of fairness.

By 1968, nobody seemed to like termination, but no one wanted to end it, either. Presidents declared it a dead policy. Congress even tried to grapple with it. At the instigation of Senator George McGovern of South Dakota, Congress went on record as guaranteeing "moral and legal obligations" to Native Americans through federal policy.[17] Some representatives and senators thought this to be an overriding of House Concurrent Resolution 108, and some Indian leaders remained hopeful. Said Earl Old Person, chairman of the Blackfeet, about the word "termination": "You have caused us to jump every time we hear this word." Old Person stressed the development of Indian reservations rather than termination. He wanted the government to stop talking about termination and instead anchor its policy on economic investment and Indian self-help. Not directly ending the termination policy was "like trying to cook a meal in your tipi when someone is standing outside trying to burn your tipi down."[18]

The Supreme Court, however, was the only American legal institution to curtail the legal meaning of termination. As soon as the Menominee termination order was implemented in 1961, the state of Wisconsin began the process of trying to regulate the Menominee lands and people. In 1962, Wisconsin decided that the Menominees were subject to all state hunting and fishing regulations. For Menominees this meant the loss of many of their remaining means of survival as well as their treaty rights. The Menominee Termination Act of 1954, argued the state of Wisconsin, ended all fishing and hunting rights guaranteed in the Wolf River Treaty of 1854, which had been signed a full century earlier by the United States and the Menominee Nation.

For Justice William O. Douglas and a majority of the Supreme Court, this was sheer folly. In *Menominee Tribe of Indians v. United States* (1968), Douglas found that termination did not mean the abrogation of existing treaty rights.[19] To end treaty rights required specific language in law to do so, and the Menominee Termination Act did not specifically override the Wolf River Treaty of 1854. Moreover, Douglas tied the termination acts to Public Law 280. In a rather novel legal twist, the Court ruled that the language in Public Law 280 (''Nothing in this section . . . shall deprive any Indian or any Indian tribe, band, or community of any right, privilege, or immunity afforded under Federal treaty, agreement, or statute with respect to hunting, trapping, or fishing or the control, licensing, or regulation thereof'') did not allow any state or federal official to reduce the treaty rights.[20] In fact, by specifically guaranteeing hunting and fishing rights and by not specifically ending the Menominees' nineteenth-century treaty with the United States, Public Law 280 doubly protected the Menominees.

For the time being, the Menominees' treaty rights were sacrosanct. The Supreme Court narrowly defined termination as the loss of federal support services, not the legal abolition of a tribe and its membership. If the state of Wisconsin wanted to abolish the Menominees' hunting and fishing rights, it would have to persuade Congress and President Nixon, an opponent of termination, to pass a new termination act specifically ending the Wolf River Treaty. The effect of the *Menominee* decision did not, however, undue the damage done by termination. The Menominees' tribal government was moribund. Of all the terminated tribes, only the Klamaths had been able to sustain any continuing tribal political manifestations.[21] *Menominee Tribe of Indians v. United States* simply put a stop to further denials of Indian rights.

The ironic coupling of termination with Public Law 280 brought to full circle the Indians' animosity regarding the direct loss to the states of their sovereignty over their reservations. They renewed their arguments that they had never been consulted about Public Law 280. For the most ardent foes of Indian sovereignty, such an undemocratic process was hard to rationalize. In addition, those states that assumed jurisdiction over Indian reservations were finding it to be an expensive process that they were not willing to fund fully, if at all. Thus a movement toward retrocession, the return of criminal and civil law functions to the federal government and tribal governments from state governments, began.

Retrocession was not debated during consideration of Public Law 280, nor was it placed in the legislation. As a legal concept, it awaited congressional approval. But some states did not even wait. Montana passed a statute in 1966 recognizing tribal consent to withdraw jurisdiction, and Nebraska tried unsuccessfully to implement retrocession on all the reservations in the state, with or without Indian consent.[22] With so many constituencies in agreement, the abolition of Public Law 280 when considering an extension of the Bill of Rights to reservations was easily available for the taking.

The Legal Origins of the Indian Bill of Rights

In 1968, the estimated number of Native Americans in the United States as counted by the BIA was approximately 760,000. Seventy percent, or approximately 530,000, lived near or on reservations. More than 85 percent of all Indians lived in eight states: Alaska, Arizona, Montana, New Mexico, North Dakota, Oklahoma, South Dakota, and Washington. The populations were concentrated even more, with 65 percent of all Native Americans residing in just four states: Alaska, Arizona, New Mexico, and Oklahoma.[23]

When Indians lived off the reservation, they were subject to the civil and criminal jurisdiction of the state in which they lived unless they were charged with one of fourteen crimes listed in the amended Major Crimes Act of 1885: murder, manslaughter, kidnapping, rape, carnal knowledge of any female not one's wife who was not yet sixteen, assault with intent to commit rape, incest, assault with intent to commit murder, assault with a dangerous weapon, assault resulting in serious bodily injury, arson, burglary, robbery, and larceny in Indian country.[24] If they committed any of these crimes on a reservation, they came under federal jurisdiction. In both state and federal courts, the Bill of Rights was a part of normal criminal and civil procedure, although many state courts were lax in their treatment of Native Americans.

Courts on Indian reservations assumed extensive powers. They had misdemeanor and some felony criminal jurisdiction when Indians were defendants and fairly wide civil jurisdiction with exclusively Native American parties and concurrent civil jurisdiction with the states when the parties were both Indian and non-Indian. Three kinds of courts existed on Indian reservations in 1968: fifty-three tribal courts, twelve Courts of Indian Offenses, and nineteen traditional courts.

The federal government recognized 435 tribes and bands in 1968, and 247 of these formally organized tribes had constitutions. Of those, 117 tribes included bill of rights provisions in their tribal constitutions, although these provisions were incomplete. Some of these tribes also had tribal courts. Although many of the tribal courts operated under a tribal constitution with bill of rights provisions, it did not matter because the

Supreme Court, beginning with *Talton v. Mayes* (1896), ruled that the federal Bill of Rights did not apply to Native American governments or courts.[25] Moreover, most of the courts on reservations did not offer rights to counsel, rights to remain silent, rights to a trial by jury, or rights of appeal to litigants. Rarely were records kept of judicial proceedings. Few judges were trained; in 1968, only five of the sixty-eight judges in tribal courts and Courts of Indian Offenses were lawyers. Clearly, Indians on reservations who were parties to disputes did not have available to them Bill of Rights protections, due process of law, or equal protection of the laws.

Beginning in 1961, two special committees, one federal and the other private, focused on Indians and the Bill of Rights. The Special Task Force on Indian Affairs recommended to the Department of the Interior that tribes protect civil liberties by passing and enforcing new ordinances. The Commission on the Rights, Liberties, and Responsibilities of the American Indian, financed by the Fund for the Republic, urged that Indian tribal governments be placed under the federal Bill of Rights because not to do so put in jeopardy ''the very assumptions on which our free society was established.''[26] The terminationists and acculturationists concluded that if they could not abolish the IRA governments, the least they could do was to have Indian institutions placed fully within constitutional constraints. The stage was set for Senator Sam Ervin of North Carolina to begin conducting hearings before his Subcommittee on Constitutional Rights into the Bill of Rights and Native Americans.

Ervin's subcommittee collected information for seven years. It sought testimony from Native Americans and interested parties. It heard stories of the abrogation of religious freedoms and personal liberty by some tribal governments. Shoddy court practices by all three kinds of Indian courts were documented. The causes of injustices on reservations were traced mainly to the judges' lack of legal training and experience, minimal resources, and the failure of the states covered by Public Law 280 to enforce laws on reservations. Field hearings by the committee recorded almost 1,100 pages of testimony.[27]

Most of those testifying explained that deprivation of rights for Indians was principally the result of improper federal and state actions. Indians were treated as slaves in local South Dakota jails; the police in Pocatello, Idaho, and Gallup, New Mexico, let Shoshonis and Navajos die in jail rather than call a doctor. The Courts of Indian Offenses cooperated with police to kidnap Indians on reservations and take them off the reservations in order to facilitate arrests. Rights to counsel were denied in state courts; not-guilty pleas were not allowed; and states included in Public Law 280, such as California, refused to allot sufficient funds for law enforcement on reservations.[28]

At the same time that the Ervin subcommittee was documenting the condition of justice on reservations, other developments concerning Indians and the Bill of Rights were taking place in state and federal courts. Before 1968, most courts adhered to the *Talton v. Mayes* decision and the Doctrine of Sovereign Immunity.[29] Tribal governments claimed that they could not be sued unless they gave permission, and they were not giving permission. Indeed, all Indian Claims commission proceedings since 1949 were predicated on this very concept. The tribal courts took the Doctrine of Sovereign Immunity seriously, but the state courts were anxious to intervene.

The full erosion of the *Talton* Doctrine came in the *Colliflower v. Garland* (1965) dispute. Madeline Colliflower, a Gros Ventre living on the Fort Belknap Reservation in Montana, was jailed after she refused to remove her cattle from tribal lands. She filed a writ of habeas corpus, and the Ninth Circuit Court of Appeals ruled that tribal courts function as federal agencies and, as such, must uphold basic concepts inherent in the Rill of Rights, such as allowing the right to counsel, the right to confront witnesses, and the right of appeal.[30] Madeline Colliflower was ordered freed from the reservation jail. *Colliflower v. Garland* was a major departure from the *Talton* rule, but as Attorney General Robert Kennedy observed in 1968 when he testified before the House Subcommittee on Indian Affairs, the *Colliflower* holding "virtually stands alone in upholding the competence of a federal court to inquire into the legality of an order of an Indian court."[31]

Although the Indian Bill of Rights adopted the legal pronouncements of the *Colliflower* decision, two other cases that resulted in holdings favorable to tribal governments probably had a greater impact on Senator Ervin and his subcommittee staff. Both disputes involved First Amendment rights. In 1953, the federal district court of New Mexico heard a dispute between the Jemez Pueblo government and six Jemez Pueblo Indians, members of several Protestant denominations who claimed that they were being religiously persecuted. The court ruled that it had no jurisdiction to hear *Toledo v. Pueblo de Jemez* (1954) because the First Amendment, as well as the Fifth and Fourteenth Amendments, did not apply to tribal governments.[32] At the same time as the Jemez

Pueblo dispute, the Navajo Tribal Council restricted religious freedoms by prohibiting the use of peyote on the Navajo Reservation. Navajos participating in Native American Church services were arrested, charged, and convicted of violating the Navajo ordinance. The Tenth Circuit Court of Appeals refused to recognize the Bill of Rights as protecting Indian religious freedom.[33]

Thus the historical and legal setting was ripe for Senator Ervin to introduce his version of an Indian Bill of Rights. Most Native Americans continued to reside on reservations or at least maintain some connection with reservation life, and Ervin's subcommittee discovered a mixed bag of Indian courts and councils. Pressures were building in the federal and state legal establishments to clarify the Indian relationship to the Bill of Rights. It seemed that only through federal legislation might this legal ambiguity be resolved, although at least one circuit court was willing to cross the lines drawn in *Talton v. Mayes*. This situation was further exacerbated by the First Amendment cases decided by the federal courts in favor of tribal institutions. Restrictions on basic religious freedoms, particularly when they might have an impact on mainstream Christian denominations, simply would not be tolerated by most Americans, regardless of the reason. As was the situation before the successful passage of the Dawes Severalty Act by Congress, assimilationists and civil rights advocates surprisingly found new ground for agreement, this time in 1968 over a new federal statute—a Bill of Rights for Native Americans.

The Indian Bill of Rights

The civil rights movement of the 1960s was powerful, successfully advocating social changes that altered the fundamental nature of American society. The challenge for Indians was to define the movement in ways that complemented their circumstances. They had to find a method to persuade Congress to abolish termination, to end relocation, and to outline Indian rights without compromising their sovereignty.[34] To shape the movement meant that Native Americans had to grasp the initiative in Congress, but few vehicles or congressional advocates were available to them. Before Indians could interpret the civil rights cause for themselves and for Congress, Senator Ervin introduced his Indian Bill of Rights.

Ervin's Subcommittee on Constitutional Rights first submitted nine bills for the Senate's consideration. Five of the proposals were aimed at correcting specific abuses in the tribal court system. Three bills offered practical protection to those who claimed that their constitutional rights had been infringed: criminal appeals from tribal courts went to federal courts for a completely new trial; a specific model code was to be created for the Courts of Indian Offenses; and the United States attorney general was charged with investigating complaints from Indians about the deprivation of their constitutional rights. All eight of the proposed new laws built on the bulwark of the Ervin subcommittee's investigations.

The ninth bill stated that "any Indian tribe in exercising its powers of local self-government shall be subject to the same limitations and restraints as those which are imposed on the Government of the United States by the United States Constitution."[35] This in essence imposed the entire Constitution on tribal councils and courts. In particular, the Bill of Rights and provisions in the Fourteenth and Fifteenth Amendments were to be extended to tribal legal institutions.

There were objections to this proposal. Many Indians thought that it posed as serious a threat to their existence as termination and relocation had. Even the Department of the Interior believed that the Ervin bill went too far. Some tribes, particularly the Pueblos, were theocracies, and they feared that the proposal would destroy their way of life. Others were concerned about the definition of tribal membership and whether it could stand up to constitutional scrutiny. Tribal resources would be severely taxed to implement the kind of judicial system mandated by the Bill of Rights, and the criminal justice system used in state and federal courts placed greater value on confrontation and punishment than did the Indian system which sought accommodation and protection of the tribe.[36]

Senator Ervin took these objections seriously and rewrote the bills. What emerged was the Indian Bill of Rights (IBR), which encompasses six sections—Titles II through VII—of the Civil Rights Act of 1968.[37] Title II was borrowed directly from the Bill of Rights and the Fourteenth Amendment. First, Indian tribes, the powers of self-government, and Indian courts are defined, and then Title II states: "No Indian tribe in exercising powers of self-government shall— . . ."[38] This is quite similar to the prohibitions of federal action in the First Amendment and state action in the Fourteenth Amendment.

Ten subsections of Title II specifically prohibit a number of Indian tribal actions against individuals. The first prohibits tribes from preventing the free exercise of religion, abridging freedom of speech or of the press, and forbidding the right to assemble or petition for a redress of grievances. These are traditionally termed First Amendment rights. The second subsection prevents tribal governments from conducting unreasonable searches and seizures or issuing warrants without probable cause. This is a Fourth Amendment right. The third subsection prohibits a person from being subjected to double jeopardy; the fourth prevents a person from being a "witness against himself"; and the fifth prevents private property from being taken for a public use without just compensation. These are Fifth Amendment rights. The sixth subsection guarantees a defendant the right to a speedy and public trial, the right to know specifically about any criminal charges, the right to be confronted by hostile witnesses, the right to subpoena witnesses, and the right to counsel at one's own expense. These are found in the Sixth Amendment, with the exception of requiring private payment for counsel. Another Sixth Amendment right is found in the tenth subsection, in which tribal governments are prevented from denying any person accused of an offense punishable by imprisonment the right to a trial by a jury of no fewer than six persons, a direct consideration of the situation in *Talton v. Mayes*. The seventh subsection prohibits excessive bail or fines, cruel and unusual punishments, and convictions for any crime for which the penalty is greater than six months in jail or a fine of $500 or both. These rights are similar to those in the Eighth Amendment. Thus Subsections 1 through 7 and 10 cover portions of the First, Fourth, Fifth, Sixth, and Eighth Amendments.

Two subsections go beyond the Bill of Rights. Subsection 8 prohibits the denial to any person of the equal protection of the laws or the deprivation of liberty or property without due process of law. These individual rights are found in the Fifth and Fourteenth Amendments. The ninth subsection does not allow tribal councils to pass bills of attainder or *ex post facto* laws. This is similar to portions of Article I, Section 9, as applied to Congress and Section 10 as applied to state legislatures.

What, then, is missing from the Indian Bill of Rights? In a sort of nineteenth-century carryover, individual Indians were accorded neither the Second Amendment right to keep and bear arms nor the Third Amendment right to be immune from the quartering of soldiers on one's property in time of peace. The Seventh Amendment, which allows the right to a jury trial in civil suits exceeding $20, was not extended to

Indian courts, principally because of the cost to tribal governments and Indian litigants.

The Ninth Amendment, the general clause retaining for the people those rights not covered in the Constitution, and the Tenth Amendment, reserving for the states or the people those powers not delegated to the federal government or prohibited to the states, were not added. Also left out was the Fifteenth Amendment, prohibiting discrimination in the right to vote based on race. The last omission was intentional, as many tribes believed that the inclusion of the Fifteenth Amendment might make it difficult for them to determine their own membership qualifications based on descent or blood quantum.

In addition to the omission of the Second, Third, Seventh, Ninth, Tenth, and Fifteenth Amendments from the Indian Bill of Rights, portions of other amendments were excluded. Noteworthy was the establishment clause of the First Amendment. The Ervin bill deleted the establishment clause after hearing extensive testimony from the theocratic Pueblo tribes of New Mexico, which argued that the establishment clause would destroy their tribal culture. The grand jury provision of the Fifth Amendment, requiring an indictment to be held to answer for a capital crime, was deleted, and the Sixth Amendment right to counsel at the expense of the government was not included. Here the Senate reasoned that tribal governments simply did not have adequate resources to retain counsel for indigent litigants, and most tribal courts did not use lawyers but instead provided tribal members for advice as needed.

The Fifth and Fourteenth Amendments were altered in the Indian Bill of Rights. The Fifth Amendment states that the federal government cannot deprive a person of "life, liberty, or property, without due process of law." Similarly, the Fourteenth Amendment stipulates that states cannot deny a person "life, liberty, or property, without due process of law; nor deny to any person within its jurisdiction the equal protection of the laws." Subsection 8 of the IBR provides that no Indian tribe can "deny to any person within its jurisdiction the equal protection of its laws or deprive any person of liberty or property without due process of law." The *life* provision of the Fifth and Fourteenth Amendments was deleted in the Indian Bill of Rights, and it is these provisions in the IBR guaranteeing due process and equal protection of the laws that have proved so contentious and litigious.

The last section of Title II also proved controversial. Initially, Ervin provided for appeals from Indian courts to federal courts to result in a trial de novo. This was challenged by many who believed that this would destroy the Indian court system. Thus in the final version, challenges to Indian government and court malfeasance were channeled through the privilege of filing a writ of habeas corpus. The only way that a person dissatisfied with his or her treatment could have that action reviewed was if the person was incarcerated. This was a narrow means of review. No language was provided in the IBR stipulating that other forms of appeal were not prohibited; they simply were not mentioned. However, if congressional intent can be measured, there is evidence that Congress preferred restricted appellate grounds.

Title II is the heart and soul of the Indian Bill of Rights. Its three sections—the definitions, the ten incorporations of selected portions of the federal Bill of Rights and other constitutional provisions, and the appellate provision using habeas corpus—constitute the only means by which individual Indians on reservations are accorded traditional Bill of Rights protections. Because Indians were not considered to be covered directly by the Constitution, the only possible vehicle available to those who wished to extend Bill of Rights concepts to Indian tribes was through congressional legislation.

Other sections of the Civil Rights Act devoted to Indian law were also path breaking. Title III directs the Secretary of the Interior to draw up a model code of justice for Congress's approval that applies to Courts of Indian Offenses. Title IV revised Public Law 280 by preventing a state from assuming criminal or civil jurisdiction over reservations without the consent of the Indian tribe or tribes in residence and by allowing any state to retrocede criminal or civil jurisdiction or both to the tribes, although no Indian consent is required. It is important to note that Public Law 280 was amended and not repealed. Title V revised the Major Crimes Act, adding assault resulting in serious bodily injury to the burgeoning list.

In the course of the Ervin committee hearings, senators discovered that the Department of the Interior had been obfuscating the ability of tribes to hire legal counsel and to know their constitutional rights and obligations. Consequently, Title VI stipulates that the Department of the Interior has ninety days to consider an Indian tribe's application for approval of a lawyer. If the agency does not make a determination, then the request is considered granted. In Title VII, the Secretary of the Interior is required to update the document "Indian Affairs, Laws and Treaties," to republish the treatise *Handbook of Federal Indian Law,* and to compile all official Interior Department opinions, published and unpublished. All these materials are to be made available to all recognized Indian tribes, bands, and groups.

The Indian Bill of Rights was indeed a significant and historic milestone for Native Americans. Its provisions changed the fundamental way in which tribes looked at themselves and their people under law. Even the partial application of the federal Bill of Rights to reservations altered forever some traditional practices. Moreover, acculturated Indians were not satisfied with a narrow interpretation of the IBR, but pressed for further expansion of this document.

Reactions to the Indian Bill of Rights

Reactions by non-Indians to the new Indian Bill of Rights were generally quite favorable. Most were surprised only to learn that Native Americans on reservations did not have access to the federal Bill of Rights, and they did not comprehend the implications of the new legislation.

The Indians' reactions were mixed. Some leaders favored the IBR. Wendall Chino, tribal chair of the Mescalero Apaches, believed that the act was an important step and hoped that the next Congress would abolish termination with a law and then adopt a major development program for the tribes. To Chino, the Indian Bill of Rights was something for all Indians to build on. The Association of American Indian Affairs also expressed its support.[39] Robert Burnette, leader of the Rosebud Sioux and former director of the National Congress of American Indians, strongly suppported the act:

> The Indian people do not have any rights. None of us are able to enter a United States district court and settle our grievances against our own elected leadership. We spent nine years fighting for the Indian Civil Rights Act. I do not think there is hardly an Indian leader today who realizes that Public Law 280 disappeared because of this act.[40]

Burnette and others mistakenly believed that Public Law 280 had been repealed.[41] One Indian present at its creation remained supportive. Helen Scheirbeck, director of the United Indians of America, believed that violations of religious freedom, the need for tribes to have their lawyers approved, and restrictions on Public Law 280 required this corrective legislation.[42]

There also was opposition. Most Native Americans were apathetic, and some were even hostile to the Indian Bill of Rights. Leslie Chapman of the Laguna Pueblo criticized the act as treating all Indians alike. She singled out Pueblo courts and Pueblo legal traditions as contrary to the IBR provisions:

> Even within the Anglo system, you can see the Constitution and due process do not necessarily yield justice. If you are concerned with how people feel and what is the effective way, what is functional in terms of the people on the reservation, the recognition of the differences is going to have to be made.[43]

The chair of the All Indian Pueblo Council, Domingo Montoya, asked for a revision of the Indian Bill of Rights to exempt all the Pueblos of New Mexico. Montoya expressed fear that the Pueblo political and legal systems would not stand up to constitutional mandates such as equal protection of the laws and "one man, one vote" reapportionment requirements.[44] Perhaps the most realistic reaction was expressed by Gerald Wilkinson, executive director of the National Indian Youth Council. To him, "every Indian is opposed to the Indian Civil Rights Act . . . until he has been screwed by his tribal council."[45]

Most of those opposing the Indian Bill of Rights felt that the act was contrary to the newly stated Indian policy of Presidents Johnson and Nixon. Opponents saw the IBR as incompatible with Indian self-determination. How, they questioned, can Indian tribes under the Indian Bill of Rights make decisions based on cultural considerations if these determinations violate Western European legal traditions? They believed that much more damage than good would result.[46] Proponents cited previous abuses and insisted that the protection afforded by the Indian Bill of Rights was justified. They also discounted the possible erosion of tribal sovereignty and treaty rights and instead saw the Indian Bill of Rights as strengthening tribal legal institutions.[47] With the first test case of the Indian Bill of Rights, this argument was demolished.

Dodge v. Nakai *(1968)*

The same week after the Indian Bill of Rights became law, the Navajo Nation was in turmoil. On April 16, 1968, the Navajo Tribal Council met to consider the operation of Dinebeiina Nahiilna Be Agiditahe, Inc.

(DNA), a nonprofit legal services corporation financed by the federal government's Office of Economic Opportunity. Earlier, the tribal chairman, Raymond Nakai, had asked the Office of Navajo Economic Opportunity (ONEO), the contracting agency with DNA, to consider rescinding its contract because Theodore Mitchell's conduct was embarrassing Nakai. Mitchell, a non-Indian, was the director of DNA. The executive director of ONEO, Peter McDonald, a political rival of Nakai and future Navajo tribal chairman, refused Nakai's request.

Frustrated, Nakai took his concerns to the council, which responded by holding two days of hearings on Mitchell and his policy of intervening in issues before Navajo agencies. Of particular concern was Mitchell's actions on behalf of several Navajos in a dispute over the operation of the Chinle school system on the reservation. The council decided not to ask that Mitchell be temporarily relieved of his duties.

Not all council members were satisfied. Annie Wauneka, a member of the council and of the Navajo Tribal Council Advisory Committee, sought out Mitchell the next day and told him that she believed he was detrimental to the Navajo people and should leave the reservation. In the next week, conditions deteriorated to the point that the Advisory Committee passed a resolution demanding Mitchell's dismissal. On July 3, Tribal Chairman Nakai told Mitchell that he had fourteen days to remain as director of DNA or the Advisory committee would take further action. Nothing happened.

On August 5, the Navajo Tribal Council Advisory Committee held a meeting in the council chambers. At the meeting were representatives of the Department of the Interior to discuss the implications of the new Indian Bill of Rights Act. Also present were DNA officials, including Theodore Mitchell. Annie Wauneka asked whether the act prevented the Navajo Nation from evicting persons from the Navajo Reservation. Durard Barnes, Acting Associate Solicitor on Indian Affairs from the Department of the Interior, asked Wauneka whether she had anyone specific in mind. She interrupted and stated that she did not, but at this point extremely loud, derisive, and insulting laughter erupted from several in the audience, the most noticeable coming from Mitchell. The next day, the meeting continued. At this meeting, Wauneka got up from her chair and confronted Mitchell. She demanded to know whether he intended to laugh at the Navajo Council again. After Mitchell said something to her, Wauneka slapped him several times and told him to leave the chambers. Mitchell left.

The next day, August 7, the Advisory committee passed a resolution requiring the immediate removal of Mitchell from the Navajo Reservation. He could then return on August 8 for a hearing about his permanent banishment. Navajo police were dispatched to remove Mitchell; he was taken from the reservation; and on August 8, there was a hearing at which Mitchell was allowed to make a statement before the Advisory committee. The committee then formally declared by a vote of 12 to 3 that Mitchell was banished from the Navajo Reservation.[48]

Two important issues surrounded the banishment order. First, Mitchell's conduct during the Advisory Committee's meeting was more than reprehensible. According to Navajo culture and tradition, disruptive, antisocial conduct is essentially a crime against the tribe. To some, including council member Annie Wauneka, such an action is tantamount to a serious felony. Second, even though Mitchell's action involved illegal personal activity on his part, *banishment* is essentially a civil action, not a criminal finding. Because it is a civil action and the restitution for any wrongful action does not require imprisonment or fines, tribal judicial institutions have jurisdiction. Banishment allows for the restoration of tribal harmony. Antisocial conduct on a reservation by an Indian or a non-Indian can be considered by tribal courts or, as in this case, a quasi-legislative/judicial tribal council advisory committee.

Mitchell brought suit in federal court to reverse his banishment. As a plaintiff, Mitchell joined eight Navajos from the board of directors of DNA and a class of indigent Navajos who used the services of DNA. One of the Navajos representing the class was John Dodge, a relative of one of Raymond Nakai's political opponents. Mitchell sued Nakai, chairman of the Navajo Tribal Council; V. Allen Adams, superintendent of the Navajo police; and Graham Holmes, area director of the Navajo Reservation for the BIA.

The district court in Arizona agreed to hear arguments regarding whether it had jurisdiction as a result of the new Indian Bill of Rights. Mitchell peppered the court with possible ways in which the federal courts could assume jurisdiction. Most were rejected, although the court did accept two of the arguments he raised. First, the district court ruled that the Navajo Tribal Council Advisory Committee's action was subject to treaty interpretation and so allowed for federal jurisdiction. Although that was sufficient, the court went further. It ruled that when Congress passed the Indian Bill of Rights, it intended to ''make substantial changes in the manner in which Indian tribes could exercise their quasi-

sovereign powers.'' This, to the court, opened the door to federal jurisdiction. The legislative intent behind the IBR was so broad that the fact
that Mitchell did not exhaust the Navajo court system before he came to
the federal court was unnecessary. Moreover, the lack of a writ of
habeas corpus was not even fully discussed.[49]

The court also dealt with the *Colliflower* precedent. The Navajos'
case presented a different situation. The Navajos never approved the
New Deal governments, so their government was based on treaties.
Consequently, the federal court in *Dodge v. Nakai* could not declare the
Navajo Advisory Committee as an extension of a federal agency in order
for jurisdiction to be attached.

After agreeing to hear the case, the court eventually issued its opinion
in March 1969. It found that under the Indian Bill of Rights, the actions
of the Navajo Nation violated Theodore Mitchell's rights by not according him due process of law (Title II, Section 202[8]) and freedom of
speech (Title II, Section 202[1]). The Advisory Committee also violated
the Indian Bill of Rights by passing what the court determined to be a bill
of attainder (Title II, Section 202[9]). The court held that the banishment
provision of the Navajo treaty with the United States in 1868 required
fundamental aspects of due process and that any action of the tribe
activating such ''a severe remedial device'' must be in full compliance
with its own Navajo tribal code.[50]

Dodge v. Nakai was a massive hemorrhage in the tribal sovereignty of
the Navajos and that of all Native American nations. Treaty rights were
attacked. Narrow jurisdictional limits placed in the Indian Bill of Rights
were expanded significantly. Restrictions placed on banishment as a
remedy for tribal governments made it almost impossible to obtain. Due
process of law would be applied to tribal courts as if it were a federal
rather than a tribal legal concept. Even Senator Sam Ervin had not
anticipated such a sweeping interpretation of his Indian Bill of Rights.[51]

■   ■   ■

''Will Capture had told his daughter a lot about old Eagle Capture, how
much he had admired his father and wanted to please him,'' writes Janet
Campbell Hale. Hale, a Coeur d'Alene, explains in her novel *The Jailing of Cecelia Capture* that Cecelia Capture, Will's daughter, is in jail
remembering a childhood event that she now understands. At the time, it

did not make sense why her father threw away all of his law books. Eagle Capture

was the one who had brought the white man's system of justice to the tribe. He believed that the key to survival was legal representation. If the Indian people had had adequate legal representation, there would have been no Little Bighorn or Wounded Knee. It wouldn't have been possible for the white-eyes to steal land and murder Indians. Legal representation was the key.[52]

Cecelia, who had shared her father's hopes, now shared her father's doubts.

This year, 1968, was indeed important to the relationship of Native Americans to the Bill of Rights. It began with the foment associated with an unpopular war and a nation attempting to come to grips with the fundamental legal and moral inconsistencies in its race relations. The assassination of America's foremost civil rights leader punctuated the nation's agony, and the political system could not contain the confusion. Another assassination brought memories of a slain president from the death of his brother. One president gave up; a new one was chosen. Indians had some cause to worry. Richard Nixon's previous federal experience was as vice president for Dwight Eisenhower, the primary advocate of termination and relocation policies in the 1950s.

For Indians in 1968, a renewed threat to tribal sovereignty appeared in the form of the Indian Bill of Rights. It was not enough for the Indian Claims Commission to continue slowly down its terminationist path or for the federal courts to make war on tribal governments. The protection afforded tribal courts and tribal councils from *Talton v. Mayes* was eliminated by Congress. Moreover, with the *Dodge v. Nakai* decision, it appeared that tribal ways of life were about to come to an end. The very nature of Indianness was at stake.

With these serious and threatening legal signs for Indians and the tempo of the times, it is not surprising that 1968 also ushered in a new way of thinking for some Native Americans. Red Power came of age. Young Native Americans, many living in the cities, the victims of relocation, proclaimed their own social movement. Northern Plains Indians in Minneapolis started a new organization, the American Indian Movement, known as AIM. Its founders, notably Dennis Banks (Chippewa), Russell Means (Sioux), Hank Adams (Assiniboine), Clyde Warrior (Ponca), and Clyde Bellecourt (Chippewa), fashioned rhetoric that appealed to the young. The activists in AIM developed a militant program that targeted corrupt tribal governments, the BIA, the FBI, the state courts and police, and irrational federal policies.

After 1968, what might the 1970s and 1980s bring? Self-determination and economic development were federal policies debated front and center. Richard Nixon offered some surprising new initiatives, and Native Americans embarked on a perilous journey with their new Bill of Rights.

John Ross (Cherokee), leader of unsuccessful resistance to Indian removal in American courts during the 1830s, in 1862. (Courtesy Nebraska State Historical Society, photo no. R539-000026/000004)

The signing of the Treaty of Fort Laramie in 1868, with Red Cloud, the Sioux and other Indian nations, and General William T. Sherman. (Courtesy National Archives, photo no. 30, 111-SC-95986)

Forced acculturation was achieved through government regulation of supplies to reservations, such as at Fort Sill on the Kiowa–Comanche–Apache Reservation, Indian Territory, 1870. (Courtesy Southwest Collection, Texas Tech University, Ernest Wallace Papers, Lubbock, Texas)

Omahas in cadet uniforms endure the Carlisle Indian Boarding School, 1880. (Courtesy National Archives, photo no. 153, 75-IP-1-10)

Internal agents of forced acculturation included the Sioux Indian police, Pine Ridge Agency, Dakota Territory, 1882. (Courtesy National Archives, photo no. 2, 75-IP-2-19)

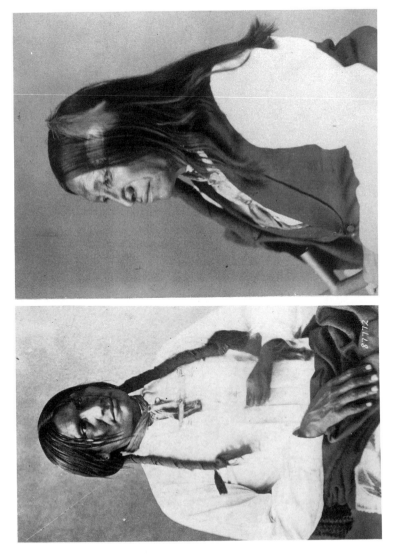

(*left*) Big Foot, leader of the Miniconjou Sioux, before his murder at Wounded Knee. (Courtesy National Archives, photo no. 90, 111-SC-87772) (*right*) Spotted Tail (Brulé Sioux) in 1872, before his murder by Crow Dog and the evolution of the important Supreme Court decision *Ex parte Crow Dog.* (Courtesy National Archives, photo no. 125, 75-ID-27)

Sioux sun dance in 1874, before its banning by the Bureau of Indian Affairs and the imposition of restrictions on Indian religious freedoms. (Courtesy National Archives, photo no. 43, 111-SC-83156)

Indian New Deal constitutionalism in action at a Kiowa council meeting, Anadarko, Oklahoma, 1940. (Courtesy National Archives, photo no. 75-N-KI0-92)

Blackfeet logging crew of the Indian Civilian Conservation Corps, Browning, Montana, early 1940s. (Courtesy Sherburne Collection, University of Montana Archives, and William Farr, University of Montana, Missoula, Montana)

Ira Hayes (Pima) as a member of the Marine Corps Paratroop School in 1943, before his heroism at Iwo Jima and his relocation to Chicago. (Courtesy National Archives, photo no. 193, 75-N-PIM-33)

(*above*) Daisy Pino (Acoma Pueblo), during job training at Brown's Cafe, Albuquerque, New Mexico, 1951. (Courtesy National Archives, photo no. 150, 75-N-ALB-6) (*below*) Relocation propaganda featured an Indian family in San Jose, California, 1957. (Courtesy National Archives, photo no. 75-N-REL-H)

Johnnie Saux (Quinault) holding a dog salmon in 1936, signaling the beginning of fishing rights disputes in the Pacific Northwest. (Courtesy National Archives, photo no. 58, 75-N-TAH-21)

(*above*) Sioux leader David Long (Crazy Horse) before the Indian Claims Commission, 1957. (Courtesy National Archives, photo no. 75-CL-91Q-3) (*below*) Peter McDonald (Navajo) before the Council of Energy Resource Tribes, 1972. (Courtesy National Archives, photo no. 75-CL-91-0-1-11)

Indian protests over the disregard of treaties and the infringement on rights culminated in the Trail of Broken Treaties, 1972. (Courtesy National Archives, photo no. 75-CL-91Q-3)

The new legal warriors, Indian lawyers, are represented by K. Kirke Kick-ingbird (Kiowa), speaking at The Museum, Texas Tech University, 1982. (Courtesy John R. Wunder, University of Nebraska–Lincoln)

Recent legal struggles concern the legalization of gambling on Indian reserva-
tions. The origins of gaming in Indian culture are represented by these four
Nuaguntit Paiutes gambling in southwestern Nevada, 1873. (Courtesy National
Archives, photo no. 67, 57-PE-71)

8

Modern Tribalism

Tobacco is an important commodity for Native Americans. It has religious, economic, and diplomatic meaning. An old Passamaquoddy story concerning the scarcity of tobacco tells much about the pitfalls in the era of modern tribalism. In this tale, a young Passamaquoddy watches his grandmother smoking, and he, too, wants to smoke. She says that she will share it as grandmothers are wont to do, but she worries. Tobacco is scarce. The grandson promises to find some new tobacco, and she tells him that the only place to do so is on an island.

The young man travels to the island, but there he confronts a woman who tries to prevent him from gathering any tobacco. A struggle ensues between the boy and the woman, and the woman changes into a crow with a bag of tobacco clutched in her claws. The boy is also transformed into a large bird, and the two fight in the air. After an extended battle, the crow/woman releases the tobacco, and the bird/boy swoops down and seizes it.

The dutiful grandson then returns with the tobacco to his grandmother, and he proudly proclaims, ''Ho'k'mi yut, t'ma'wei kwuskwe'sul [My grandmother, here is the tobacco].'' His grandmother wisely answers, ''Ndege'k'ma'jehan [You'd better go your way]; k'dunlogo'kw [she will be after you].''[1]

The caution of the Passamaquoddy should be considered in any assessment of the recent legal history of Native Americans. After fifty years of ''reforms'' that resulted in new tribal governments, termination, and a partial Bill of Rights, Native Americans of the 1970s to 1990s assess their legal rights carefully and view reality cautiously. After 1968, rights were defined by the Indian Bill of Rights for individual Native Americans on reservations. During the 1970s, an expansion of these rights at the cost of tribal governmental power and tribal culture was at first encouraged in the federal appellate courts and then discouraged through the Supreme Court's decision in *Santa Clara Pueblo v. Martinez* (1978).

Similarly, the pressures of the 1970s brought with them a new pantribalism that combined a variety of Indian legal grievances. Foremost was the drive to recapture lands previously lost. The federal government cooperated in what it perceived as moderate requests, such as the return of Blue Lake to the Taos Pueblo or the return of lands to several New England tribes, including the Passamaquoddy. For some Indians, however, this was too little, too slowly. They then couched their rhetoric and actions in militant gestures and extralegal activities, such as the occupation of Alcatraz Island. The federal legal establishment responded with greater force, such as at the Wounded Knee II shoot-out.

Confrontations over land and cultural issues moved from the rocks and the fields to the courtrooms and the legislative chambers. Indian lawyers and Indian law took over the fight. The modern era witnessed the end of termination and the Indian Claims Commission and the implementation of a new policy of restoring federal recognition of terminated tribes. Laws were formulated reflecting the new federal policy of self-determination for all Indians.

Power and Primacy:
Struggling to Achieve Limited Sovereignty
and Unlimited Rights,
1969–1978

Three completely different men held presidential power during the 1970s. Richard Nixon was a man that Americans either hated or loved, and he eventually was forced to resign in 1974 in the wake of Watergate scandals. He was not replaced by his initial vice president, Spiro Agnew, as Agnew had his own legal troubles. He, too, resigned to plead *nolo contendere* to a variety of charges stemming from bribery and influence peddling. Gerald Ford, a plodding, lackluster long-term congressman from Michigan, briefly replaced Nixon until he was defeated in 1978 by Jimmy Carter, an obscure former governor of Georgia. The Nixon–Ford presidency generated factional fighting and unrest in the country, with Nixon's impeachment hearings captivating a national audience.

For Native Americans, there was much irony in the state of power and governance in the United States during the 1970s. So often accused of not being able to govern, of being indecisive, of choosing corrupt leaders, and of dividing into polarized political factions, Indian tribes now were witnessing at the federal level a full-scale realization of the often mythical behavior that non-Indians accused them of perpetuating on their reservations.

Amazingly, out of this political atmosphere emerged programs and laws of great benefit to Native Americans. Modern tribalism was born, and its emergence came as a result of the strong presidential power and support that Richard Nixon devoted to it. Jimmy Carter followed Nixon's Indian legal policies with additional important federal initiatives. Significant legal transformations of Native American rights came in the 1970s, a period of change that rivaled the Indian New Deal.

A Decade of the Indian Bill of Rights

When the Indian Bill of Rights was proposed, Domingo Montoya, chairman of the All Indian Pueblo Council, requested that his people be exempted from it. He testified before Congress, ''Our deep concern over the Indian Bill of Rights stems from our fear that it will destroy the traditions—and in doing so greatly weaken our governments.''[2] Montoya proved to be prophetic, for the first decade of the Indian Bill of Rights in action constituted a direct challenge to the basic right of Indian governments to govern themselves. In particular, the first interpretations of the Indian Bill of Rights by federal courts were inconsistent with congressional intent and broadly expanded federal powers, at the expense of tribal governments and traditions.[3]

During the congressional hearings held in the early 1960s, senators were quick to point out that Congress primarily wanted to help strengthen tribal institutions while protecting individual Native American rights. The author of the bill, Senator San Ervin, on more than one occasion reiterated his desire to maintain and preserve tribal and cultural practices. This was not an easy task. It defied the statutory construction given to the diversity of culture to be taken into consideration. Thus it remained for the courts to respond to individual situations. Congress certainly had not anticipated the judicial results.[4]

Within five years of the passage of the Indian Bill of Rights, amendments were introduced into Congress to stem the loss of tribal rights. The National Tribal Chairman Association proposed an amendment that would require Indian tribes or bands to ratify the Indian Bill of Rights before it could be applied to them. The association hoped that this would slow the steady loss of tribal governmental power.[5] But Congress did not adopt any changes, even though many Indians believed that the Indian Bill of Rights needed alteration.

Native Americans were troubled by the implementation of the Indian Bill of Rights because federal courts using the IBR started stripping away the powers of tribal governments. The initial litigation concerned political rights of individual Indians and non-Indians. This took a variety of forms. Clarification of rights included free speech, reapportionment, tribal-council candidacy, suffrage, election irregularities, and tribal membership, all challenging the very foundations of Native American tribal governments.

When the Indian Bill of Rights was adopted, few thought that free-speech issues would be the first requiring court intervention. Vine Deloria, Jr., then executive director of the National Congress of American Indians, testified at hearings that ''for the most part, you have these rights naturally as your customs. . . . [N]o one prevents you from free speech. No one prevents you from assembly, because Indian society is simply that way.''[6]

Yet almost immediately, free-speech issues were raised on the Navajo Reservation when an unpopular legal services director publicly humiliated members of the Navajo Tribal Council. When they responded with an appropriate cultural punishment, banishment, the director went to the federal courts, which intervened on behalf of the attorney.[7] Seven years later, a different free-speech issue arose on the Pine Ridge Reservation. Several employees of the Oglala Sioux tribe's community health program were terminated because they had criticized the tribal government. When the fired workers went to federal court, they argued that their free-speech rights under the Indian Bill of Rights had been violated. The court recognized these rights, but it ordered the employees to exhaust all tribal remedies before coming to the federal courts.[8]

Reapportionment was immediately controversial. ''One man, one vote'' was definitely not an Indian concept, but the federal courts insisted on its application to Indian political institutions. In *White Eagle v. One Feather* (1973), members of the Standing Rock Sioux tribe sued on the basis of the equal protection clause of the Indian Bill of Rights and *Baker v. Carr* (1962) to change the apportionment of their tribal council. The federal court granted their challenge, incorrectly stating that eliminating malapportioned districts would not force ''an alien culture with strange procedures on this tribe.''[9] Similarly, three Crow Creek Sioux tribal members sued their tribal council. Here, in *Daly v. United States* (1973), the federal circuit court held that districts had to have close apportionments based on population, not Crow Creek's eligible voters.[10] In regard to reapportionment, the federal courts were anxious to intervene, determining that provisions for equal protection of the laws in the Indian Bill of Rights were to be interpreted as if tribal governments were the same as state or federal governments.

Who could run for the tribal councils was also of great concern. In its tribal constitution, the Rosebud Sioux prohibited employees of its public health clinic from running for the tribal council. This was challenged, and although the federal circuit court in *Luxon v. Rosebud Sioux Tribe* (1972) ruled that this was a proper question to be considered under the Indian Bill of Rights, it did not decide the issue.[11] When write-in candidates beat the incumbents on the Goshute Tribal Business Council, the council refused to give up power. The write-ins sued and lost, because the federal court decided that the tribal government had not passed an ordinance allowing write-in ballots to be counted.[12] Blood-quantum requirements for candidacy were upheld as being within the range of tribal cultural interests, and the Potawatomis in Wisconsin were allowed to exclude women as candidates.[13]

Who could vote was litigated. Two cases involved challenges by eighteen-year-olds who argued that they were entitled to vote because of the Twenty-sixth Amendment to the Constitution. Courts found against incorporating the Twenty-sixth Amendment into the Indian Bill of Rights, and so tribal constitutional age limitations of twenty-one years were retained. The reasoning for this determination was that federal courts should not force a new class of voters on Indians who may not want them to vote. Equal protection provisions in the Indian Bill of Rights would not be construed to apply.[14] It is particularly difficult to square the reapportionment cases with the franchise cases.

A number of legal confrontations arose over election irregularities.

When the Sisseton–Wahpeton Sioux Tribal Council allowed a felon to run for tribal secretary and sanctioned an unauthorized recount of ballots for tribal treasurer and a second election for secretary, the federal courts were asked to intervene, and they ruled partially in favor of the disappointed office seekers.[15] When Russell Means complained that Richard Wilson was fixing the election for tribal chairman on the Pine Ridge Indian Reservation, federal courts were less sympathetic and ruled that Means had not exhausted his tribal remedies.[16] But when Winnebagos in Nebraska challenged the election of their tribal council, they were successful in obtaining a temporary injunction to prevent council members from being seated.[17]

By 1974, the federal courts were becoming less anxious to intervene against tribal governments, in part because of the Supreme Court's decision in *Morton v. Mancari* (1974), in which it upheld the constitutionality of a federal statute providing for employment preference for Native Americans in the BIA.[18] Here federal equal protection guarantees allowed for positive discrimination on behalf of a trust-protected political class. It was therefore easier for the lower federal courts to be more cautious. To protect tribal governments, either they could rule that equal protection or due process provisions in the Indian Bill of Rights must be defined on Indian terms, or they could insist on exhausting all remedies in Native American societies before approaching the federal courts.

This latter device was widely used. When several members of the Cheyenne River Sioux tribe defaulted on a tribal loan for their ranching business and had their grazing unit lease canceled, they sued in federal court. Because the defaulters did not attend informal tribal hearings, their failure to exhaust all tribal remedies imperiled their federal court suit.[19] Conversely, when Henry Cobell, a Blackfoot, tried to regain custody of his two children from their mother after he had won custody in a Montana court divorce, he found tribal courts blocking his way. The Blackfoot Tribal Court issued a restraining order to prevent Cobell from taking his children. He then filed for a writ of habeas corpus, which was granted by the federal circuit court. It found in favor of Cobell because the tribal code specifically prohibited tribal courts from taking jurisdiction over divorce and child-custody actions. The tribal court judge admitted that he was testing his own jurisdiction, and the federal court would have nothing to do with it. Here the exhaustion of tribal remedies was interpreted strictly.[20]

Consequently, the first decade of interpretation of the Indian Bill of Rights was eventful, with the IBR inviting instant litigation. Tribal governments were under constant attack. Numerous political issues were heard before federal courts, which were simply not prepared to grapple with them. In this sorting-out time, three trends emerged. First, the federal courts tried to apply federal constitutional standards, but this proved unsatisfactory. Eventually, the judges resorted to delays, during which they forced the tribal governments to put in place local methods of deciding disputes. Second, Indians began to develop their own concepts of due process and equal protection, and they had some limited success with educating the federal courts. This led to the recognition of tribal custom as a legitimate way to resolve delicate political rights issues. And third, Native Americans needed some confidence in their own tribal institutions without the constant threat of federal intervention or litigation. Indians and tribal governments were not always capable of sustaining regular battles in court over basic concepts of tribal life, mainly because of the expense.

By the end of the decade, one remaining political right had yet to be fully litigated, and it was a fundamental right. Who could be a member of a tribe, and how should this be determined? Before the Indian Bill of Rights was instituted, those decisions were up to the tribe. Now the federal courts and even the Supreme Court entered the fray. The Pueblos had been right from the beginning about their fears concerning the Indian Bill of Rights, and it would be they who eventually became the center of the most important decision interpreting the act.

Santa Clara Pueblo v. Martinez *(1978)*

In 1941 Julia, a full-blooded member of the Santa Clara Pueblo, married a Navajo man, Miles Martinez. They had several children. The couple resided at the New Mexico pueblo, and the children were raised on the reservation, where they were brought up in the Santa Clara Pueblo culture and spoke the pueblo's Tewa language. The Santa Clara Pueblo contains 1,200 members on the reservation and 150 off the reservation. In addition, approximately 200 nonmembers live on the reservation.

Before the Martinez marriage, in 1939, the Santa Clara Pueblo Tribal Council passed an ordinance stipulating that "children born of marriages between female members of the Santa Clara Pueblo and nonmembers shall not be members of the Santa Clara Pueblo." A second ordinance provided that "children born of marriages between male members of the Santa Clara Pueblo and non-members shall be members of the Santa Clara Pueblo."[21] Julia Martinez's husband was prevented from becoming a naturalized member of the Santa Clara Pueblo because it was the desire of the Santa Clara Pueblo to maintain a patrilineal society through its membership. The result was that the Martinez children, with a full-blood Indian mother and father, were denied membership in the Santa Clara Pueblo.

To rectify this situation, Julia Martinez, on behalf of her daughter Audrey, sued the governor of the Santa Clara Pueblo, Lucario Padilla, and the Pueblo itself, asking for injunctive relief against the tribe to prevent it from denying membership to her children. It was not simply a matter of joining the tribe. At stake were medical services, educational opportunities, and other federal benefits. Martinez alleged that Title II of the Indian Bill of Rights was controlling because her children were being denied the equal protection of the laws.

At the federal district court, the judge ruled that he had jurisdiction. He stipulated that by its very existence, the Indian Bill of Rights waived all tribal sovereign immunity and gave federal courts the right to hear cases alleging rights infringements on reservations. This was an extremely broad interpretation of the habeas corpus clause of the Indian Bill of Rights and congressional intent. Having granted jurisdiction, at the subsequent trial the district court held against the Martinez children. The judge ruled that the ordinance of the Santa Clara Pueblo, albeit of recent composition, reflected long-standing cultural traditions of patriarchy and constituted the tribe's own self-definition. Given the importance of these laws, the Indian Bill of Rights could not be used to strike them down.[22]

The Martinez family next appealed to the Tenth Circuit Court of Appeals, which agreed with the district court's contention that the federal courts had jurisdiction, but it disagreed with the district court's findings on the merits. Accordingly, the appellate court ruled in favor of Julia Martinez, reasoning that the protections guaranteed in the Fourteenth Amendment must be applied. Here was a classification by the Santa Clara Pueblo based on sex, and as such it was "presumptively invidious." There was no compelling tribal interest to retain such discrimination.[23] In essence, the appellate court applied federal equal protection standards to the Indian Bill of Rights.

After the Tenth Circuit expanded the Indian Bill of Rights even further, the Santa Clara Pueblo had no recourse but to appeal to the Supreme Court. The case was argued in November 1977, and the opinion of the Court was issued in March 1978. The justices broke down in a 7 to 1 opinion, with Justice Thurgood Marshall for the majority and Justice Byron White dissenting. Marshall's opinion, backing the tribe, put a screeching halt to the appellate court's expansion of the Indian Bill of Rights and, at best, made further use of the IBR difficult. *Santa Clara Pueblo v. Martinez* proved to be a landmark opinion.

Justice Marshall framed the issues in such a way as to recast the Indian Bill of Rights. He first established the semisovereign nature of tribal governments, noting that Indian tribes were "distinct, independent political communities retaining their original natural rights' in matters of local self-government."[24] Then he offered the unqualified statement that Indian tribal governments "have power to make their own substantive law in internal matters."[25] He concluded by reinforcing his notions of sovereignty by describing the tribes as remaining "quasi-sovereign nations which, by government structure, culture, and source of sovereignty are in many ways foreign to the constitutional institutions of the Federal and State Governments."[26]

If tribal governments have extensive powers, then how can they be sued? Marshall concluded that they cannot be sued without a specific congressional mandate. "Indian tribes," he wrote, "have long been recognized as possessing the common-law immunity from suit traditionally enjoyed by sovereign powers."[27] The only way they can be sued is if Congress has "unequivocally expressed" itself on the issue, and in the Indian Bill of Rights that did not happen. The federal courts had erred in allowing the plethora of suits against tribal governments.

There was a way out, however, for the Martinez family. Under the habeas corpus provisions of the Indian Bill of Rights, an officer of the tribe could be sued, and the family had enjoined the Santa Clara Pueblo governor. Neither he nor any other officer was immune from suit. But the Martinez family still lost because the only remedy available under the IBR, according to the majority, is the writ of habeas corpus, and it did not apply in this instance. Marshall went on to close the door to any court's hearing any cause of action under the Indian Bill of Rights other than for writs of habeas corpus. Injunctive relief or any other remedy was not explicitly supplied under the act, and Congress expressed its legislative intent against a broad range of remedies. Thus if the Indian

Bill of Rights were to be changed, it would be up to Congress and not the courts to do so.

Marshall noted that the Indian Bill of Rights contains two competing purposes. On the one hand, individual tribal members are to have their rights strengthened against their tribal governments, but this is to be done at the same time as tribal self-determination is encouraged. How is one, then, to interpret the act when the two are in conflict, as they were in *Santa Clara Pueblo v. Martinez?* Marshall concluded that the answer was in the IBR itself. Because the Indian Bill of Rights did not incorporate *all* the federal Bill of Rights, Congress must have meant that if individual rights conflicted with tribal self-determination, the tribal government should prevail. This is a most bizarre form of reasoning that defies rational explanation.

Moreover, Marshall went even further. In a slap at the federal appellate courts, Marshall stated that when "Congress seeks to promote dual objectives in a single statute [in this situation, with the Indian Bill of Rights], courts must be more than usually hesitant to infer from its silence a cause of action that, while serving one legislative purpose, will disserve the other."[28] That is, the courts were to be restrained. This was a congressional province, and not civil rights business as usual.

Thus *Santa Clara Pueblo v. Martinez* was a significant legal determination. It halted the federal courts' extravagant and expansive interpretations of the Indian Bill of Rights, and it stemmed the erosion of power away from tribal governments. No longer would interpretations of the Indian Bill of Rights be made almost exclusively outside Native American legal institutions.[29]

Indian Extralegal Activities

In addition to the landmark decision in *Santa Clara Pueblo v. Martinez,* the end of the 1960s and the beginning of the 1970s brought with them some changes in Native American perspectives on their rights. Numerous pan-Indian organizations were formed in the 1960s, principally to work to redress past federal policies. Most of the organizations sought to make legal changes within the established political system, but younger Native Americans were impatient and not sure that this could be done thoroughly or quickly enough. They formed new groups, the first being the National Indian Youth Council (NIYC), which grew out of a meeting in Gallup, New Mexico. They demanded Indian control over their culture, education, religion, and politics. ''Our viewpoint,'' according to their Statement of Policy, ''based in a tribal perspective, realizes literally that the Indian problem is the white man, and, further, realizes that poverty, educational drop-out, unemployment, etc., reflect only symptoms of a social-contact situation that is directed at unilateral cultural extinction.'' Such a conclusion demanded direct action.

Why had so many Indians reached the point of confrontation and extralegal activity? First, Native Americans were in difficult straits. In 1970, the average annual income for Indians on or near reservations was $2,800; those living off reservations fared somewhat better, with $4,700. Unemployment on the reservations averaged 50 percent, with some reservations at levels as high as 90 percent. Over half of all Native American families lived in substandard housing. Furthermore, life expectancy for Indians living on reservations was forty-four years, twenty years less than for non-Indians.

Second, although relocation policies had been mitigated, the flow to urban areas, especially by young Indians, continued. This meant significant change for Indian tribes and families. Between 1953 and 1972, more than 100,000 Native Americans left the reservation for the city. In 1940, only 13 percent of all Indians lived in metropolitan areas, but by 1980, approximately 52 percent of all Native Americans lived in urban areas. Thus many children were being reared off the reservation in their formative years.[30]

Third, Indians in cities and young Indians generally were defining their place in tribal society, and in this process they were seeking a means of placing Native Americans in the civil rights movement. This led to conflict with federal authorities and conservative tribal governments. The best-known organization to emerge from this process was AIM, the American Indian Movement.

The first major event of the Indian civil rights cause was the occupation of Alcatraz Island in San Francisco Bay on November 9, 1969, by a group of students from Bay Area colleges. They were removed the next day by federal marshals, but on November 20, they returned to stay for what turned out to be nearly two years. The occupation's leaders included Richard Oakes (Mohawk), Earl Livermore (Blackfoot), Dennis Hastings (Omaha), and John Trudell (Santee Sioux). The abandoned prison had been occupied earlier, in 1964, by a small group of Sioux, but they were quickly removed by federal officials. The later group tried for four years to gain title to the island through the federal courts, but they eventually lost. Alcatraz had no water or food, and the Indian occupants lacked strong leadership and financial support. Finally, after nearly two years, on June 11, 1971, the federal government moved in and retook Alcatraz.[31]

Alcatraz was just a beginning. In 1972, AIM held a convention in Minnesota and demanded that the Chippewas be given control over their lakes. AIM also led more than 1,000 Indians to Gordon, Nebraska, to protest the murder, by whites, of Raymond Yellow Thunder. Protests were also held for other Native Americans killed, including Richard Oakes, leader of the Alcatraz occupation, who was slain by a prison guard in California. For the summer of 1972, AIM planned the Trail of Broken Treaties caravan to Washington, D.C., which began in Minneapolis and culminated in the takeover of the BIA building in Washington. AIM demanded the return of full sovereignty to all tribes; the repeal of the Resolution of 1871, which prevented future treaties between Indian nations and the United States; and the abolition of state court jurisdictions over Indian reservations.

Militant Indians also took on the tribal governments. The leaders of AIM, especially Russell Means and Dennis Banks, believed that many tribal governments were corrupt and should be overthrown. Throughout the country, tribal leaders on reservations were challenged. This movement accelerated in 1973 into civil war on the Pine Ridge Reservation in South Dakota. The previous year, Richard Wilson, a conservative and a BIA backer, defeated Gerald One Feather, a traditionalist, for tribal chairman. Wilson hated AIM and tried to destroy it on his reservation. He even sent the Pine Ridge tribal police to help Rapid City officials control AIM demonstrators.

Russell Means and Dennis Banks decided to provoke Wilson by moving onto the reservation. Federal marshals came also, and on February 28, 1973, AIM took over a small settlement at Wounded Knee— basically a church and a trading post named after the historic massacre and creek—and declared its independence from the United States. AIM demanded that Richard Wilson be removed as tribal chairman and that the Lakota treaties receive a new federal review, separate from the proceedings of the Indian Claims Commission. At first, the federal government refused to discuss the issues that AIM raised, and it brought in the FBI. Several planned assaults were considered, but negotiations were begun and allowed to proceed. After two months of sporadic gunfire, AIM agreed to end the occupation if federal lawyers would work with the Oglala Sioux factions on treaty disputes. AIM left, and the attorneys came. Finally, all parties agreed that only Congress could act on the matter.

The 1970s proved that Native Americans would no longer be subservient, that they were prepared to take to the streets or to the hills to fight for what they believed were their rights. Moreover, Indians were willing to take on not only the federal and state governments, but their own tribal governments as well. Into this atmosphere of confrontation came a significant shift in federal policy, to self-determination.

Self-Determination

The doctrine of self-determination has been a fundamental feature of Native American legal history. Since the ratification of the Constitution, most of the legal considerations of Indian self-determination have centered on its limits. Self-determination was featured in the first Supreme Court cases to consider Indian issues, most notably the Cherokee cases, and its definition has been crucial to all subsequent discussions of Indian sovereignty, civil rights, and government. Nevertheless, self-determination for Native Americans took on new meanings in the last half of the twentieth century.

The first president to embrace a new meaning for Indian self-determination was Lyndon Johnson, in his speech to Congress on March 6, 1968. Johnson's speech, entitled "The Forgotten American," was the first address given by a president to Congress specifically concerning Native Americans. In his talk, Johnson's definition of Indian self-determination was more image than substance. Paternalism continued to be a basic part of the Indian–federal government relationship. Johnson did suggest that economic activities were the key to self-determination as a policy, a concept that proved to be enduring. Only twenty-five days before announcing his retirement from politics, he urged Congress to come up with the programs necessary to make Indian self-determination work.[32]

The definition of self-determination for many Indians was much different from that of Lyndon Johnson. To them, self-determination meant legal and political sovereignty. Treaties would be honored literally; homelands would be returned and protected; and the trust relationship would be maintained. Some even argued that independence and separate nationhood were intrinsic to self-determination, rather than limited economic development supervised by the federal government.

To the surprise of many, Richard Nixon endorsed a much more revolutionary framework for Indian self-determination than his predecessor had. Nixon was much more resolved to define Indian self-determination specifically and to make it official federal Indian policy. In his 1968 campaign, Nixon issued a policy statement embracing the nebulous concepts of Indian self-determination; then as president, Nixon sent a special message to Congress on July 8, 1970. This message—crafted in part by the Commissioner of Indian Affairs, Louis R. Bruce, a Mohawk–Sioux from Pine Ridge, and LaDonna Harris, a Comanche and the wife of Democratic Senator Fred Harris of Oklahoma— contained a specific program for implementing Indian self-determination policies. Bruce wanted Indian tribes to contract with the federal government, rather than the BIA, for tribal services. Nixon also urged the abolition of termination and paternalism by the federal government. He defined *self-determination* as encouraging "the Indian's sense of autonomy without threatening his sense of community."[33] Nixon wanted Congress to abolish termination with a specific resolution; to require Indian control of Indian programs, especially education; and to set up an independent source of legal representation for Native Americans so they could protect their own interests.

No non-Indian politician had attacked termination before with such vehemence. What motivated President Nixon is unclear. Earlier, his nominee for Secretary of the Interior, Governor Walter Hickel of Alaska, had caused some controversy because he had delivered speeches in which he advocated termination. But Nixon himself identified with the underdog, and so perhaps he had some legitimate empathy for Native Americans. Then, too, self-determination as a theory of self-help fit basic Republican party values and Nixon's own personal views. The fact that Nixon's religion was Quakerism, long associated with advocating Native American rights, and that he counted as one of the two most influential men in his life his old college football coach, Wallace Newman, who was an Indian, probably had some influence as well.[34] Whatever the case, Nixon strongly pressed Congress to abolish termination and accept self-determination:

> Because termination is morally and legally unacceptable, because it produces bad practical results, and because the mere threat of termination

tends to discourage greater self-sufficiency among Indian groups, I am asking the Congress to pass a new Concurrent Resolution which would expressly renounce, repudiate and repeal the termination policy. . . . [S]elf-determination among the Indian people can and must be encouraged without the threat of eventual termination. In my view, in fact, that is the only way that self-determination can be effectively fostered.[35]

Congress delayed. It was not ready to consider new definitions of Indian self-determination. In the past, it had allowed occasional limited aspects of political and economic self-determination, but in the end Congress always placed checks against the full development of political or economic autonomy by tribes. Then, Watergate intervened and Nixon resigned, and before Watergate, Wounded Knee II forced the resignation of Commissioner Bruce. But the ideas embraced by the Nixon administration did not go away.

In 1975, Congress finally passed portions of the Nixon-Bruce agenda. The Indian Self-Determination and Education Act[36] and the American Indian Policy Review Commission Act[37] were then endorsed by President Gerald Ford. These two pieces of legislation made Indian self-determination the cornerstone of modern federal Indian policies, and they still are operative.

The Indian Self-Determination and Education Act consists of three major sections. The basic federal policy toward native peoples is the first part of the statute. Congress admitted that federal paternalism retarded rather than enhanced life for Native Americans and that Indians "will never surrender their desire to control their relationships both among themselves and with non-Indian governments, organizations, and persons."[38] Moveover, Congress promised to work in all areas toward Indian self-determination, especially in education, while maintaining the trust relationship. Indians were also given preference in hiring for all federal government contracts relating to Indian tribes.

The second legislation passed in 1975, when Congress created the American Indian Policy Review Commission, was more controversial. The commission was to assess the historical and legal developments regarding the United States and American Indians. Congress noted in the preamble that no such assessment had been made since the 1928 Meriam Report.[39]

The commission was composed of six members of Congress and five Indians (three from reservations of recognized tribes, one from an urban area, and one from a nonrecognized tribe). James Abourezk, Democratic senator from South Dakota, chaired the commission, with Lloyd Meeds, Democratic congressman from Washington, as his vice chair. The executive director of the commission was Ernest Stevens (Oneida), and K. Kirke Kickingbird (Kiowa) served as general counsel. Thirty-one of the thirty-three members of the task force were Indian.[40]

A number of duties were statutorily assigned to the commission. It was to study treaties, statutes, judicial opinions, and executive orders to determine the current federal–Indian relationship. It was to pay particular attention to legal ways to protect Indian resources and to recognize Indian groups previously not recognized. It was to investigate ways that Indians could be represented at the national level through elective means, and in an obvious reference to problems with the Indian Bill of Rights, Congress charged the commission with the "consideration of alternative methods to strengthen tribal governments so that the tribes might fully represent their members and, at the same time, guarantee the fundamental rights of individual Indians." Finally, the commission was to recommend laws that would help Indian tribes realize their political and legal self-determination.[41]

This was a significant alteration in the normal congressional view of Indian self-determination, and this departure from previous policy was made all the more clear when the commission issued its Final Report in 1977. It found, among its 206 recommendations, that tribal governments should be treated as local and state governments, with responsibilities outside constitutional interference. The commission endorsed the tribal governments' full sovereignty over reservations. This included taxation of both Indians and non-Indians, criminal and civil jurisdiction in tribal courts over both Indians and non-Indians, Native American government control of reservation watersheds, and the full exercise of treaty rights over and against the states and non-Indian localities in regard to traditional hunting and fishing activities.[42]

Vice Chairman Meeds could not support this expansion of the definition of self-determination, and he said so. He published an extensive minority report, in which he wrote: "The doctrine of inherent tribal sovereignty adopted by the majority report ignores the historical reality that American Indian tribes lost their sovereignty through discovery, conquest, cession, treaties, statutes, and history."[43] Meeds, of course, was incorrect. True, Indian sovereignty had been held in abeyance, but it was not because of international or American law. It was because of force. If anything, history was on the side of the resurgence of sovereignty within the doctrine of Indian self-determination.

Indian self-determination was now the law of the land, and it remained for it to be practically defined. Many Indians throughout the 1970s and 1980s tried to shape self-determination in ways comparable to those outlined by the American Indian Policy Review Commission. Richard Real Bird, tribal chairman of the Crow Nation, noted in 1989 that to his people, "sovereignty" is a strong word, like "miracle," "freedom," and "liberty." The Crow have sovereign rights not granted by Montana, the United States, or the BIA. Real Bird defined self-determination as the right for the Crows to speak in their own language, to practice their own religion, and to maintain hegemony over their own territory. But Real Bird also recognized that self-determination meant economic sovereignty as well. Eighty-five percent of his people on the Crow Reservation were unemployed, and 95 percent were on some form of assistance. Economic sovereignty as a form of self-determination, Real Bird believed, must occur first.[44]

At the national level, the first steps toward fleshing out the rights of American Indians under the policy of self-determination came with the return of homelands, the federal recognition of "new" tribes and the restoration of terminated tribes, and the protection of Indian children, religion, and natural resources.

Homelands Restored

A basic part of the Indian self-determination policy revolution, strangely enough, centered on the return of lands to certain tribes after long-standing grievances. These grievances were not settled by the Indian Claims Commission, which, by the early 1970s, was on its deathbed. The Taos Pueblo were the first native peoples to benefit from this significant change.

To the people of the Taos Pueblo, Blue Lake is a sacred place in which life began. Located in north central New Mexico, Blue Lake is the spiritual and economic focus of Pueblo life. It was a part of the Taos Pueblos' homeland until 1906, when the federal government placed it and 48,000 other acres in the Kit Carson National Forest. Taos Pueblos were outraged when non-Indians were allowed access to their sacred lake, and they demanded the return of both it and their homelands.

Beginning in the 1950s, the Indian Claims Commission took up the

Taos Pueblo case, and in 1965 it declared that Blue Lake had wrongfully been taken. However, instead of encouraging Congress to return the lake, the ICC offered the Taos Pueblo $10 million and several thousand acres near the lake. The Pueblo leadership refused the offer. Throughout the 1960s, the Taos Pueblo people demanded the return of Blue Lake and their lands, but the Department of Agriculture refused to do so for fear that other tribes would also want their lands back.[45]

Richard Nixon knew of the Blue Lake controversy and decided to assign John Erlichman, his chief domestic adviser, to do something about it. Erlichman, in turn, asked his aide Bobbie Kilberg to work on it, and she obtained the help of LaDonna Harris. Together they planned a media blitz and legislation to return Blue Lake and 48,000 acres to the Taos Pueblo. The struggle to free Blue Lake focused on the Senate, which for the first time overruled the chair of the Interior Committee—in this case, Henry Jackson of Washington. Jackson vehemently opposed the return of lands to Native Americans. Once the bill reached the floor of the Senate, no one knew how the vote would go until Barry Goldwater, conservative senator from Arizona, stood up to have his say. To a hushed audience, Goldwater stated his position, "I've watched the Taos Struggle and I think we should even turn Phoenix over to the Indians, so the least we can do is support this measure."[46] The leader of the Pueblos, the cacique, stood up and lifted the cane that his people had received from Abraham Lincoln. The Senate broke into thunderous applause, and the Blue Lake measure passed. Later when the Blue Lake legislation–signing ceremony was held, Richard Nixon offered some impromptu words, and everyone wept. The return of Blue Lake in December 1970 was a momentous event for the Taos Indians and all Native Americans.[47]

Issuing executive orders and forcing legislation out of the Senate Interior Committee were perhaps easy victories when compared with other pressing claims of Native Americans, especially those of Alaska's Indians. Native Alaskans number more than 80,000 Indians, Inuits (also termed Eskimos),[48] and Aleuts. Because of Alaska's unusual history, its Native Americans had never been the subject of treaties with the United States, and this legal weakness was made abundantly clear in the Supreme Court's decision in *Tee-Hit-Ton Indians v. United States* (1955). Thus most of Alaska's Indians had no legally recognized tribal sovereignty or trust status with the United States. They held a unique legal position among American native peoples.[49]

In 1951, Native Alaskans submitted claims to the Indian Claims Commission for the hundreds of thousands of acres they had lost to the state of Alaska and the federal government. Almost 300,000 acres, they pointed out, were managed by the Bureau of Land Management. Although the Alaskans outlined the lands in question in their ICC claim, the federal government in 1966 allowed, without consultation, portions of the Brooks Range to be opened for oil and gas exploration. Alaska's natives raised objections, and the Secretary of the Interior placed a freeze on leasing until the Native Alaskans' claims were settled.

The freeze worked. Congress and the federal government had to act in order for the oil companies to move ahead with their development plans. Legislation was prepared, and in 1971 the Alaska Native Claims Settlement Act (ANCSA) became law. Although the oil companies were pleased, the mining, fishing, and hunting corporations and organizations were not. Native Americans in Alaska claimed 60 million acres, but they settled for 40 million acres of the public domain.[50]

Under the Alaska Native Claims Settlement Act, 40 million acres of Alaska were set aside, and $962.5 million was appropriated for their distribution and management by native regional corporations. This was the largest settlement in land and dollars ever made to Native Americans by the United States. ANCSA had held the Nixon administration to its promise to move Native Americans toward self-sufficiency while maintaining the federal trust relationship.

Not everything was positive. There were potential problems, the most important one being the terms for the operation of the thirteen regional native corporations. First, the native corporations were closed for twenty years; then after 1991, individual native shareholders could sell their stock. This situation was reminiscent of the Dawes Severalty Act in practice. Stock was a twentieth-century name for allotment. The potential for Native Alaskans to lose all that they had gained at the end of twenty years was great. Soon before the 1991 deadline, however, in 1988, an amendment to ANCSA was adopted, prohibiting the sale or transfer of stock unless the regional corporation approved.[51] This will provide an additional hurdle, but it does not prevent the abolition of the corporation and, with it, control over lands and trust funds.

Other deficiencies have been identified. The act puts a cap on oil and mineral royalties for Native Alaskans of no more than $500 million or no more than 2 percent, without time limitations. This is in effect a minuscule share. The corporations require each native village to organize into a unit. The corporation itself can decide whether it wants to recognize a native village as such. If the decision is not accepted, the litigants can appeal to the Alaska Native Claims Appeal Board (ANCAB), a federal agency, which has the final say in the matter. This will lead to constant bickering about what constitutes a village, and Native Alaskans who do not reside in the villages are generally not eligible under ANCSA.[52]

The impact of the land returns has been mixed. Some were genuine acts of compassion and respect by Congress and a president anxious to rectify a wrong, such as the return of Blue Lake to the Taos Pueblo. Others, such as the Alaska Native Claims Settlement, were compromises that, albeit noble in purpose, have forced nonnative institutions on native peoples with damaging results, not unlike those of the Indian New Deal and termination. Other returns of homelands were forced by the courts and administrative agencies as a result of new interpretations of old laws.

Recognition and Restoration

"Recognition" and "restoration" have specific meanings in the legal history of Native Americans, and they are relatively recent concepts. Both are derived from reactions to laws passed in the earlier part of the twentieth century. They were also the subject of significant investigation by the American Indian Policy Review Commission, whose recommendations did much to encourage further legal clarifications.[53]

Recognition is the formal means by which the United States officially acknowledges the existence of a Native American group or tribe. It is analogous to the official standing of foreign nations with the United States government. Recognition of Native American nations has important ramifications, for it usually means that federal programs are extended to the members of the recognized entity. Moreover, the Indian Bill of Rights and the IRA can also be applied to a newly recognized tribe. Recognition can be achieved through an act of Congress, an executive order, or the federal courts.

Restoration is the process through which a tribe must go in order to be recognized after it has been terminated. It is a direct response to the termination acts passed for specific tribes in the 1950s and 1960s. This designation comes only from Congress. It has been from the 1970s to the present that the recognition and restoration movements have been most active.

Before the 1970s, recognition was not a particularly contentious concept. Congress had not determined what a tribe was, and the federal courts had not clarified the situation. The Department of the Interior recognized tribes on a haphazard case-by-case basis. Recognition, however, became an important legal concept in the 1970s, enough to provoke a number of court decisions and a federal regulation from the Department of the Interior specifically defining a tribe.

The steps leading to the codification of recognition and the administrative determination of tribal status were somewhat convoluted. Probably the most important by-product of the Indian Claims Commission was the discovery of numerous groups of tribes in the United States that had not been officially recognized by the federal government and that did not have specific treaty relationships with the United States. This happened because the enabling legislation that established the ICC made it easy for any Indian group, recognized or unrecognized, to file a claim. Once a claim was filed, it was only a matter of time before the ICC or the courts faced the issue of whether the group was in fact a band or a tribe. The ICC was purposely created to assist in the termination of tribes, but instead it ironically led to an increase in the number of tribes defined under federal law.[54]

In addition to the ICC's playing an important role in forcing the executive branch to codify tribal recognition, the courts also issued important rulings that accelerated the process. In 1974, the Supreme Court held in *Morton v. Mancari* that tribes were a political, not a racial, classification.[55] Non-Indian employees of the BIA had brought a class action against Department of the Interior regulations and federal statutes giving Indians preference in BIA jobs. It was a direct challenge to the new Indian self-determination policy as framed by President Nixon and Congress. Indian self-determination was upheld, but this case had perhaps a greater impact on recognition decisions. Groups of Indians that intermarried with other racial groups were not in danger of having their children's mixed-blood status used against them. Under American law, tribe was not synonymous with race.

A second case that caused a rash of recognition petitions was *Joint Tribal Council of the Passamaquoddy Tribe et al. v. Morton et al.* (1975), decided by the First Circuit Court of Appeals.[56] The Passamaquoddys have lived in Maine for at least 3,000 years. Their relationship with the United States began with a treaty of alliance during the American Revolution in 1777. Although they fought against the British, they were not included at the Treaty of Paris negotiations. Their land base on the St. Croix River was eroded by a 1794 treaty with the state of Massachusetts (Maine was then a part of Massachusetts), by which they ceded most of their land rights in return for two reservations on the Canadian border, approximately 27,000 acres. For almost 200 years, the Passamaquoddys have watched the state of Maine take portions of their land base, whittling it down to 17,100 acres without proper compensation or consultation.[57]

The Passamaquoddys decided to sue for their lands based on the Indian Trade and Intercourse Act of 1790, which specifically prohibits the sale of Indian lands without a public treaty with the United States. The Passamaquoddys never signed a treaty with the United States to cede their land claims, and so their treaty with Massachusetts and the subsequent actions by Maine were in essence void. Furthermore, the Passamaquoddys were not recognized by the United States as an Indian tribe. In 1892, Maine's supreme court decided that the Passamaquoddys were not a tribe,[58] despite their two reservations and some 1,300 members, 800 of whom lived on the reservations.

The First Circuit Court judges ruled that the Passamaquoddys were indeed a tribe and therefore were entitled to trust benefits and protections under the United States. In reaching this conclusion, the court considered whether the tribe was racially and culturally intact and whether a community of Indians had dealt with federal, state, or local governments over time. In both cases, the Passamaquoddys passed scrutiny. What all this meant was that the Indian Trade and Intercourse Act of 1790 prevailed and that title to much of Maine theoretically had yet to be extracted from Maine's Indians.[59]

Congress quickly got into the act. The Passamaquoddys agreed not to press claims for occupied lands, but they wanted cash and unoccupied lands. The Maine Indian Claims Settlement Act was passed by Congress on October 10, 1980, and signed by a lame duck President Jimmy Carter on December 12.[60] The act set up a $27 million trust fund and a land acquisitions fund of over $54 million for the Passamaquoddys and their neighbors, the Penobscots and the Houlton Band of Maliseets. Each has purchased lands in Maine with the funds. The settlement with the Passamaquoddys opened the door to other Indian groups in New England, none of which had signed treaties with the federal government and none of which had been recognized. Most were successful in obtaining recognition and a claims settlement.[61]

Another court ordered the Department of the Interior to make administrative law. After the *Passamaquoddy* decision, the department received nearly forty petitions for recognition, but decided to stonewall the onslaught. One tribe caught in this was the Stillaguamish, and they went to court. The federal court in the District of Columbia ordered the Department of the Interior to make a decision on the petition within thirty days, thereby forcing the executive to draft guidelines for the determination of recognition.[62]

Finally, the Department of the Interior issued temporary administrative regulations in 1978 and permanent ones in 1980. In its "Procedures for Establishing that an American Indian Group Exists as an Indian Tribe," three requirements were made for Indians to be recognized as a tribe. First, the group had to be ethnically and culturally identifiable. There were no racial requirements, which allowed for acculturation and intermarriage. Second, the group was required to have lived in "a substantially continuous tribal existence," which allowed for a brief slippage of tribal identification. And third, the group must have acted autonomously from the colonial era to the present.[63]

There were limitations. All three requirements had to be met in order for a tribal group to become a recognized tribe. The regulations applied only to the continental United States, and so Native Hawaiians were left out. Of significance was the limitation on the dynamics of tribal development. No "splinter groups, political factions, communities or groups of any character which separate from the main body of a tribe currently acknowledged as being an Indian tribe by the Department" can gain recognition "unless it can be clearly established that the group has functioned throughout history until the present as an autonomous Indian tribal entity."[64] This provision had two effects: It locked tribes into a static political stage of development, and it did not allow for the federal government's past mistakes in which separate tribes, sometimes national enemies, were forced to live on a common reservation and participate under a government not sanctioned by one of the tribes.

Although these provisions may seem onerous, they turned out to be much more flexible than those adopted by the federal courts making decisions immediately after their publication. In *United States v. Washington* (1979), a Pacific Northwest federal district court held that to be recognized, tribes had to live as continuously separate, distinct, and cohesive Indian cultural or political communities; exercise sovereignty over their members or any territory; and maintain organized tribal political structures.[65] In Massachusetts, the First Circuit Court ruled in *Mashpee Tribe v. New Seabury Corp. et al.* (1979) that a four-part test had to be met in order for the Mashpees to be recognized. They were required to be of the same race, united in a community throughout time, governed by continuous leadership and one government, and concentrated in a defined territory. The Mashpees failed all four parts.[66] In both the Pacific Northwest and Mashpee situations, the courts decided to ignore the new federal regulations and issue their own opinions, which were much narrower and more restrictive. The message seemed to be that after 1978, if Indian groups denied rights through recognition sought relief, they had to turn to the executive and legislative branches rather than the judiciary. Such was the tone and achievement of Indian self-determination and the new tribalism.

Restoration was another matter that required executive and legislative attention, and its evolution effectively ended termination with Congress's having to be placed on record as having made a fundamental error. Termination caused a number of legal problems for tribes. The tribal governments no longer had authority over their members, who lost their tax exemptions, federal program benefits, and tribal lands. Termination also meant the loss of recognition. Tribes that had been terminated were de-recognized in their individual termination acts, and they were prevented from applying for recognition by the Department of the Interior recognition regulations.[67]

The first tribe to be terminated, the Menominees of Wisconsin, was also the first to be restored. Termination had been extremely destructive to the Menominees. Before termination, the Menominee trust fund contained nearly $10.5 million. The tribe used the interest from these funds to establish and maintain health care facilities, educational buildings, a lumber mill, and a power plant. But by 1972, the fund had been depleted to $58,795.[68]

With termination, the Menominees' tribal properties were placed under the control of a corporation, Menominee Enterprises, Inc. (MEI). MEI held 90 percent of the reservation's taxable property, but even so it could not manage the new state tax burdens. MEI also became a bank holding company that the Menominees had little power to influence. In order to pay taxes and reduce them at the same time, MEI began to liquidate its landholdings. It sold them to individual non-Indians and to companies that planned to build resorts. In the fall of 1971, the Menominees managed to take control of the MEI board and, in the process, began to ask for the restoration of their tribal status and federal recognition. Several members of Wisconsin's congressional delegation introduced the Menominee Restoration Act in April 1972.

In December 1973, the Menominee Restoration Act became law.[69] The act was blunt; it specifically repealed the Menominee Termination Act of 1954, emphasizing that no treaty rights had been altered, including fishing and hunting rights. The Supreme Court had already assured the Menominees of this, but Congress reinforced the guarantee. A Menominee Restoration Committee was placed in charge of the Menominees until their tribal government could be reactivated, and the membership roll, previously closed, was reopened.

The Menominee Restoration Act recognized the Menominees as a tribe once again, and it returned to the Menominees a number of rights that they had lost. But the act placed under tribal control only those lands currently held by MEI. Consequently, the Menominee Restoration Act is known not only for what it did, but also for what it did not do. The act did not provide for the return of lost reservation lands; it did not reimburse the tribe for its lost trust funds; and it did not offer a settlement for the loss of lands or the human suffering caused to the Menominees by termination.[70]

Still, the Menominee Restoration Act was important, as it signaled the end of termination to other terminated tribes, and they decided to seek restoration. Termination therefore was effectively abolished, even if Congress could not bring itself to repeal Concurrent Resolution 108.

After the Menominee Restoration Act was passed, Congress established six criteria that terminated tribes must satisfy in order to regain recognition. All six have to be met for a tribe to be restored. First, the terminated tribal members or their ancestors have to have been gathered in an identifiable community. Second, the community has to be located near the former reservation. To satisfy these two requirements, a tribe essentially demonstrates its tremendous resistance to termination. Third, a self-governing organization has to be functioning for the tribe, and fourth, the tribe has to demonstrate that its language is being used and its culture is being observed. Tribes that want to qualify for these two criteria set up committees and hold powwows. They reinforce the processes of educating their young in the ways of the tribe. By conforming to these two criteria, tribes essentially prepare to assume tribal status. Fifth, restoration can be granted if tribes can prove that their socio-economic conditions have deteriorated after termination, and sixth, tribes have to prove that they are poorer than their non-Indian rural neighbors. These last two criteria are simply the ratification of the obvious, but they are also the most expensive to prove. Consultants are hired by the terminated tribes to survey their members to obtain the information needed to prove these criteria. As in the recognition process and with termination itself, the BIA has refused to help the terminated tribes.[71]

With the restoration criteria clearly articulated, terminated tribes started immediately to regain their rights and retain their tribalism. Table 2 shows that most of the 109 tribes terminated have been restored. Each restoration has been unique. Some, such as the Klamath, took a great deal of time, not resulting in restoration until 1986. Some tribes obtained federal land for a new reservation. The California Rancherias were restored by lawsuits because their termination acts required the Department of the Interior to complete certain water projects that had never been finished. They successfully argued that they were never really terminated.

The last tribe to be restored were the Northern Ponca, and that came about in 1990. Their restoration was held up by Congressman Douglas Bereuter of Nebraska until the Northern Ponca Restoration Act was amended to prevent the Poncas from ever being given a reservation. Termination sentiments are still alive in Congress. Yet to obtain restoration are the Mixed Blood Utes in Utah and the Catawba in South Carolina. Both are involved in legal actions that must be resolved before they can request restoration status from Congress.[72]

Although most of the terminated tribes have been restored, it has not been without cost. Restoration has taken both years and a substantial investment, in the hundreds of thousands of dollars. All could have been avoided. But in many ways, it was a miracle that those Native American peoples who were ''liquidated'' survived at all; restoration truly is a tribute to their ability to persist. Wrote one anonymous Northern Ponca on her survey sheet returned to the Ponca Restoration Committee:

> It's a shame we have to be beyond hope or desperate to the point of death before our cry for help is heard. My cry is for you to hear the plea of those

TABLE 2.   Restoration of Terminated Indian Tribes

Tribe or band	State	Termination Act	Termination Date	Restoration Act
Menominee	Wisconsin	1954	1961	1973
Klamath	Oregon	1954	1961	1986
Western Oregon	Oregon	1954	1956	Restored*
Alabama–Coushatta	Texas	1954	1955	1987
Mixed Blood Utes	Utah	1954	1961	Pending
Southern Paiute	Utah	1954	1957	1980
Lower Lake Rancheria	California	1956	1956	Restored[†]
Wyandotte	Oklahoma	1956	1959	1978
Peoria	Oklahoma	1956	1959	1978
Ottawa	Oklahoma	1957	1959	1978
Coyote Valley Ranch	California	1957	1957	Restored[†]
California Rancherias[‡]	California	1958	1961–1970	Restored[†]
Catawba	South Carolina	1959	1962	Pending
Northern Ponca	Nebraska	1962	1966	1990

*Includes 61 bands and tribes. All Oregon tribes and bands requesting restoration have been restored; many were restored in groups as confederated tribes. Restoration took place over a period of years.

[†]All California Rancherias requesting restoration have been restored. Restoration took place over a period of years.

[‡]Includes 37 or 38 rancherias.

Source: Beth Ritter Knoche, "Termination, Self-Determination and Restoration: The Northern Ponca Case" (M.A. thesis, University of Nebraska–Lincoln, 1990), pp. 58, 79–82.

who can still do for themselves if assisted by learning, for those who are not yet born. There is so much one can do with the knowledge someone cares.[73]

Last Gasps of the Termination Era: The ICC and Public Law 280

With the advent of recognition and restoration, it appeared that the termination era was drawing to a close. Termination was found to be wanting by just about everyone, including the states. It had not worked. Two of the foundations of termination also were restrained during the era of modern tribalism, the Indian Claims Commission and the expansion of Public Law 280.

From 1969 to 1978, the Indian Claims Commission continued to function. It was essentially a cleanup operation. Cases pending after the expiration of the ICC were afforded the courtesy of being sent to the Court of Claims. During its last years, the ICC was concerned primarily with correcting its previous abuses, including the gratuitous offset.

The gratuitous offset was defined as the costs not specifically promised in treaties or agreements that were borne by the federal government on behalf of Indian tribes. Usually in the form of services or food, these expenditures were subtracted from any funds the ICC designated for Native Americans.[74] The Department of Justice abused the gratuitous offset in cases decided in the 1950s and 1960s, and it worked. Indeed, some tribes refused federal projects in the 1960s because they remembered the "hidden price tags" that the Justice Department lawyers brought forward at ICC hearings.[75] Both the ICC commissioners and Congress knew about these abuses by the executive.

In 1969, the Indian Claims Commission decided to do something about the gratuitous offset. In the case of *Delaware Tribe of Indians v. United States* (1969), the ICC concluded that the Department of Justice could not subtract the value of a reservation supposedly made as a gift to the Delawares.[76] The commissioners reasoned that the United States gained more than it lost. Peace and the assurance that the Delawares were confined to a specific place had significant value to the fearful national psyche. This was a far cry from the ICC's previous decisions.[77]

In the *Delaware* case, however, the ICC tried at least to salvage something for the government. The commissioners ruled that offsets amounting to more than 5 percent of the award for additional considerations of land would be allowed. The small offsets—those that irritated the general public and Congress—were ignored, but the large offsets could still be litigated. This was a direct slap at the Supreme Court, which had ruled one year earlier in *Peoria Tribe of Indians of Oklahoma et al. v. United States* (1968) that the United States was liable for the interest on funds it agreed to hold for the Peorias.[78]

Congress got into the act at this point. The House held hearings on the offset controversy, and in 1974 it passed a bill out of committee that amended the Indian Claims Commission Act. Under the amendment, the ICC could not consider offsets for food or provisions. The amendment was directly aimed at the ICC's consideration of the claim to the Black Hills by the Sioux. This dispute was litigated for over fifty years, and the ICC, nearing the end of its statutory life, sought to reach a conclusion. The value of the land in dispute was $17.1 million in 1877, the date that the ICC decided to use, but the federal government submitted as a gratuitous offset a bill for nearly $25 million. The result would be no Black Hills claim. Most of the offset was in the form of rations and provisions given to the Lakotas to keep them from starving.[79]

Before the House could pass its bill, the Senate attached it to the ICC's operating budget for fiscal year 1975, and the House then agreed. The change was signed into law on October 27, 1974.[80] This was perhaps the most important modification in the laws under which the ICC operated during its entire existence. Yet it came very late. Since it was not retroactive, it did little to correct the previous unfair—and some would argue immoral—offsets granted.

The Indian Claims Commission ended in 1978. It exhausted Congress. It did not finish its work, and it complicated things more than it helped. The process of claims resolution became extremely tedious, and it made tribes and the United States into adversaries, making compromises impossible to obtain. From 1951 until 1974, the ICC authorized awards of approximately $534 million. Because of the decision to evaluate land at nineteenth-century prices, the awards amounted to only 2 percent of the total land value. Of the $534 million, $53 million alone went for lawyers' fees. One-half of the claims cases were dismissed. Had Native Americans brought these actions as non-Indians under the Fifth Amendment, they would have received much greater compensation. In fact, the ICC was created so that Native American nations would lose any incentive to remain tribes. But the Indian Claims Commission virtually strangled itself with its own rules attempting to prohibit fair compensation. Like termination, in the end the ICC could not even follow through with its own mandate.[81]

Just as the ICC floundered about in the 1970s, so too did Public Law 280, although its life span was merely put on hold. In the Indian Bill of Rights of 1968, Congress placed significant limits on the states' ability to assume criminal and civil jurisdiction over Indian reservations. No longer could the states simply take control of law enforcement on reservations without the permission, in a referendum, of the resident Indians. In addition, states that had taken jurisdiction could return it to the federal government, although Indians did not have the right to demand retrocession, nor did they have the right to prevent retrocession.[82]

The requirement of tribal consent for further applications of Public Law 280 was significant, as it represented a statutory beginning for the implementation of Indian self-determination. Moreover, because tribes had to give permission to states to assume jurisdiction over their reservations, states were no longer agents of termination, but simply providers of needed services to the Indian nation consumer.

Retrocession, or the return by states of criminal and civil jurisdiction over Indian reservations to the federal government, was much more controversial. The lack of tribal consent or even formal consultation was the root of most of the controversies. The absence of Indian voices had allowed the states to request partial retrocessions that sometimes were very detrimental to the reservations.

The biggest controversy was in Nebraska. Although this was a mandatory state under Public Law 280, it had refused to provide even rudimentary services to its reservations. After the revisions to Public Law 280, Nebraska asked to retrocede its jurisdiction over both the Omaha Reservation and the adjacent Winnebago Reservation. The Secretary of the Interior granted the retrocession of the Omaha Reservation but not the Winnebago, because the Winnebagos did not want retrocession. The state of Nebraska was not happy with this arrangement, so its legislature tried to rescind the proposal in its entirety. The courts refused to let Nebraska take this action. Meanwhile, law enforcement on both reservations was in limbo.[83]

Public Law 280, therefore, is still alive, unlike the Indian Claims Commission, although it has been limited in its effectiveness, and its basic thrust had been reshaped. Public Law 280 is now a partner, to some degree, in Indian self-determination; no longer is it the great threat to tribal sovereignty that it once was.

■   ■   ■

By the end of the 1970s, a peace pipe had been figuratively passed between Native Americans and many institutions of the United States. The tobacco was lit. An eventful decade promised a new era of partnership based on mutual respect. Modern tribalism had achieved some successes. The Indian Bill of Rights had evolved to a point that it no longer threatened to destroy tribal governments. Some homelands had been restored, and the federal government and courts seemed to hold a new appreciation for past treaties. Tribes that had been victims of termination had been restored, and recognition procedures had been improved. Public Law 280 had been defanged, and fortunately the Indian Claims Commission had expired.

Nevertheless, Native Americans entered the 1980s at some risk. Modern tribalism also meant some difficult losses. If self-determination had not been official federal policy and fewer Indian nations had been prepared, the possibility of disaster would have been great. Native American rights, freshly won, soon were in retreat, caused by a new proacculturation majority on the Supreme Court. The admonitions of the Passamaquoddy grandmother remained, ''Ndege'k'ma'jehan [You'd better go your way]; k'dunlogo'kw [she will be after you].''

9

Rights in Retreat

Duane Big Eagle, an Osage Sioux poet from Oklahoma, writes haunting poetry about Native American life. In ''Flowers of Winter: Four Songs,'' he sings the ''Song of the Newborn'':

> I come from the valley of endings,
> there is no place to go but onward.
> Speak you someone
> Who knows what to say
> When they come in a dream
> Masked desperate angels
> From the valley of a massacre.[1]

For the newborn Native American, life is full of perils in the twentieth century. Modern tribalism addresses those perils, particularly the threats to economic rights, Indian families, culture, and religious freedom.

For Native Americans of the 1980s and 1990s, the masked angels are visible. Their legal rights as distinctive peoples are vulnerable. A decade earlier, the emergence of modern tribalism meant much for Indian national harmony; difficult times were easier to accept. A spillover effect resulted in the passage of several laws near the end of the decade that attempted to guarantee Indian rights. These laws concerned the protection of Indian religions and Indian children. Congress also provided new mechanisms for economic development predicated on the notion that Indian self-determination required the federal government to help tribes become economically viable. Indeed, self-determination was the catch-word for all Indian programs.

By the mid-1980s, funding for federal programs had drastically de-creased, and reservation Indians were in serious trouble. Modern tribal-ism also provoked reactions and exercises in power. The end of that era once again witnessed new attacks on Indian legal claims, which arrived in various forms, from state resistance to enforcement of the Indian Child Welfare Act to the Supreme Court's undercutting guarantees of First Amendment rights to practitioners of the Native American Church.

The 1990s nevertheless includes new hopes and a regeneration of Native American determination to prevail. A new civil rights movement is developing, this time with Native Americans defining their own des-tiny, seeking their own order, and demanding their own harmony.

Retrenchment and Racism: Refining and Maintaining Limited Rights in an Era of Restraint, 1979–1990

The progress and protection of Indian rights in the 1970s quickly came to a halt in the 1980s. Indian rights, like national civil rights concerns, no longer was a priority with Congress and particularly the executive branch. This was a conservative era with conservative leadership. Ronald Reagan and George Bush seldom discussed Indian rights; cer-tainly there was not the kind of dialogue that the nation had heard during the Johnson, Nixon, and Carter years. Indifference amd sometimes con-tempt rather than respect for native peoples characterized leadership. During the early Reagan years, James Watt was appointed to be Secre-tary of the Interior, and he favored a federal policy reminiscent of that of the 1870s and 1950s.

The Supreme Court's composition changed from a liberal court dur-ing the 1960s to first a moderate court and then a conservative court during the 1970s and 1980s. Liberal, moderate, and conservative labels do not, however, necessarily reflect pro- or anti-Indian legal positions in case law. For example, Richard Nixon was a conservative president, but his favorable view of Indian self-determination allowed him to be espe-cially helpful to Native Americans in their efforts to recapture sover-eignty initiatives. Conversely, Thurgood Marshall is generally de-

scribed as a liberal Supreme Court justice, and yet he wrote opinions in
the 1980s that were destructive to Indian rights.

The Supreme Court lost little time before turning back Native American water, hunting, and fishing rights; extending state jurisdiction over
Indian reservations; and preventing Indians from exercising First
Amendment rights. Justice Marshall's retirement seemed to leave Indian
rights issues in the hands of Justice Antonin Scalia, and his initial
ventures into the world of Indian law were extremely hostile toward
Native American legal positions. Sovereignty and individual and collective rights once again hang in the balance. Scalia, perhaps the Court's
most dogmatic conservative, seems ideologically comfortable with the
Henry Dawes and Arthur Watkins approach to Indian law. Thus Indian
legal strategy must necessarily turn to Congress to hold off the Supreme
Court's 1990s forced acculturation charge.

Hunting, Fishing, and Water Rights

Throughout the 1970s and 1980s, there was much litigation regarding
Indian hunting, fishing, and water rights. This was caused by several
factors. First, Native Americans became much more cognizant of ways
to protect their rights. The first Indian law case text, *Law and the
American Indian: Readings, Notes, and Cases,* by Monroe E. Price,
was published in 1973.[2] Indian law was taught for the first time at a
number of law schools. Programs were set up for Native American
students so that they could go to law schools. Representing individual
tribal members and class actions, Legal Services, part of President Johnson's War on Poverty, took on long-established local laws and traditions
that discriminated against American Indians. Indians themselves formed
cooperative legal rights organizations, such as the Native American
Rights Fund, headquartered in Boulder, Colorado. In essence, Native
Americans frequently found themselves in court.

At the same time that modern tribalism supported the vigorous entrance of Native Americans into the legal system, the states were more
aggressively pursuing reservation resources and monitoring Indian
rights both off and on reservations. The population push to the West
created tremendous demands for water, and the states portrayed Indian
water rights as delaying development. The states and the federal government also became concerned about conservation. Non-Indian governments tried to regulate hunting and fishing on the grounds that it was

necessary to preserve animals and fish from extinction, and these attempts conflicted with long-standing Indian treaty rights.

It is well to point out that state intervention over and against Indian hunting and fishing rights both on and off reservations is founded on the notion of a constitutionally protected inherent police power left to the states. This, however, contradicts Indian treaty provisions and the will of Congress as expressed in Public Law 280, the Indian Bill of Rights, and Indian self-determination policies.

Many Indian treaties contain specific provisions protecting Indian fishing and hunting practices. For example, in 1854 several Pacific Northwest bands and tribes, including the Puyallup and Nisqually, signed the Treaty of Medicine Creek, of which Article 3 states that "the right of taking fish, at all usual and accustomed grounds and stations, is further secured to said Indians in common with all citizens of the Territory."[3] Congress reinforced such treaty rights with Public Law 280, which stipulates in the section conferring criminal jurisdiction over Indian lands to the states that the states are not to "deprive any Indian or any Indian tribe, band, or community of any right, privilege, or immunity afforded under Federal treaty, agreement, or statute with respect to hunting, trapping, or fishing, or the control, licensing, or regulation thereof."[4] If Public Law 280's language was not comprehensive enough, fifteen years later Congress once again reiterated the protection of Indian treaty rights. In the Indian Bill of Rights, under Title IV, "Jurisdiction over Criminal and Civil Actions," Congress listed the exact protections offered in Public Law 280. Indian hunting, fishing, and water rights were to be held against the world.

These acts, however, did not stop the states from trying to assert jurisdiction over fishing and hunting rights and the Supreme Court from eroding those rights. For 125 years, disputes have been brought before American courts concerning Indian fishing rights, and until 1968, Native Anerican treaty provisions prevailed. But that all changed with Justice William O. Douglas's opinion in *Puyallup Tribe, Inc. v. Department of Game* (1968).[5] In this case, the Treaty of Medicine Creek was interpreted to allow state regulation of off-reservation nets used by Indians, ostensibly because the treaty did not prevent it. Almost as if to ensure extensive future litigation, the majority opinion did not find whether the Washington State regulation was "reasonable and necessary." Subsequent litigation attempted to define Indian fishing rights in terms of the amount of catch, types of equipment, and place of access, but the conclusion was that states could now regulate federally protected Indian fishing rights.[6]

By taking aggressive action, the states regulated the off-reservation fishing of Indians who did not have specific treaty rights, and they supervised Indian fishing if there was a clear threat to conservation that necessitated state intervention to protect the basic Indian treaty right. But the states were not through. They wanted it all. In 1983, New Mexico went to the Supreme Court hoping to be able to extend its state hunting and fishing regulations to the Mescalero Apache Reservation. The Court would not allow this, although it tried to assess the impact of New Mexico's regulations against the need for conservation. The Court's message was that if the state could clearly show that drastic conservation was necessary, then it might be able to regulate fishing on Indian reservations.[7]

Generally, hunting rights were not separated from fishing rights, but sometimes there were distinct differences. Both the states and the federal government became involved in hunting regulations. As they did with fishing rights, the states aggressively tried to prevent tribes from exercising their hunting rights guaranteed by treaties. One of the first modern conflicts over hunting rights came in Oregon when the state attempted to stop Klamaths who had been terminated from exercising their hunting rights on their former reservation's forests. Litigation worsened the situation. Some Klamaths were protected because they had not voluntarily terminated, whereas those who had willingly resigned their tribal memberships were not allowed to keep their treaty rights.[8]

Tribes also tried to regulate hunting rights on their own reservations. In Montana, the Crows prohibited hunting and fishing on their reservation by non-tribal members. The state of Montana also tried to assert its power over the reservation by regulating non-Indians. Tribal sovereignty took another blow when the Supreme Court in *Montana v. United States* (1981) held that the tribes cannot regulate riverbeds on reservations, only the states can; furthermore, the tribes may regulate non-Indians only on reservation land and only if these non-Indians are threatening the tribe's political, economic, or health survival. Such limitations are tantamount to the Indians' not being able to regulate non-Indians on their reservations.[9]

The federal government also entered into hunting regulation conflicts because of the Endangered Species Act of 1973[10] and the Eagle Protection Act of 1940.[11] The Eagle Protection Act was amended in 1962 to prohibit the killing of golden eagles as well as bald eagles and to allow Indians who wished to take eagles to apply for an eagle-hunting license with the Department of the Interior.[12] These laws were challenged by Dwight Dion, Sr., a Yankton Sioux, who was convicted of having shot four bald eagles on the Yankton Sioux Reservation. The fundamental question was whether Native Americans had the right to hunt bald or golden eagles on their own reservations for noncommercial purposes. This right, argued Dion, came from treaties, the First Amendment, and the Indian Bill of Rights' free-exercise clause. Eagle feathers are an important part of traditional Indian religions, and so enforcing the federal laws amounted to restricting the right to the free exercise of religion.

The courts were confused. The federal district court convicted Dion of having violated both the Endangered Species Act and the Eagle Protection Act. The appellate court reversed the convictions based on the Endangered Species Act, but allowed the others to stand. The Supreme Court reversed the appellate court and did great damage to Indian treaty rights. In *United States v. Dion* (1986), Justice Thurgood Marshall wrote an opinion that dealt a serious blow to Indian treaty rights and Indian civil rights.[13] Marshall held that the congressional legislation passed—the Endangered Species Act and the Eagle Protection Act—supercedes the Yankton Sioux Treaty of 1858. Consequently, Indians no longer have unrestricted rights to hunt eagles on their own reservations. Marshall refused to consider the religious implications of his holding, but the reality was that federal needs to preserve eagles took precedence over Indian needs to practice their religion.

Thus by 1990, Indian fishing and hunting rights were in a state of retreat. It is now clear that conservation can be used as a concept to overrule treaty rights and possibly even rights generally guaranteed in the Bill of Rights for Native Anericans. Reservations are not sanctuaries. The state and federal government can regulate them under certain conditions, which amounts to a direct attack by the states on Indian self-determination. Moreover, *Montana v. United States* made it extremely difficult for tribal governments to regulate their own reservations.

By the late 1970s, Indian water rights were on a stronger legal footing than fishing and hunting rights were, but they also have been weakened during the past twenty years. The courts during the 1970s and 1980s were filled with disputes challenging Indian water rights, the *Winters* Doctrine, and the *Arizona v. California* (1963) precedent, the legal underpinning essential to Native American nations to compete successfully for scarce water resources. Slowly, federal and state judiciaries have whittled away at previously defined Indian rights. The first major modification of the *Winters* Doctrine came in 1975 when in its own state courts, New Mexico sought to define Indian water rights to the San Juan River. It won, and the litigation is continuing. State courts are notoriously hostile to Indian water rights, and the Navajos—who have yet to receive their fair share of scarce water in the Four Corners area—must litigate in several state courts as a result.[14]

Modern tribalism gave Indian tribal governments the ability to fight states and private users for water, and it did not come too soon. During the past twenty years, the water issues that the courts litigated included the determination of which court system has primary jurisdiction over Indian water rights, the quantity and quality of water reserved for Native Americans, the uses of this reserved water, the sale of water reserved by Indians to non-Indians, and the extent of tribal authority over water resource management. All these issues tested the fundamental sovereignty of tribes and the strength of Indian self-determination policies.[15]

A dual system of jurisdiction over Indian water rights was accepted by the Supreme Court beginning in the 1970s. The Court ruled that the quantity of water reserved for Native Americans could be determined by hostile state courts. During the Reagan administration, the urge for the Court to return powers to the states went so far as to overrule Indian groundwater rights that had only recently been determined in *Cappaert v. United States* (1976).[16] The Wyoming Supreme Court in 1988 held that Shoshonis and Arapahoes could claim acreage that could be irrigated for agricultural and domestic use only. They could not claim water for economic development or for sale to non-Indians, and they held no claim to groundwater under the Wind River Reservation.[17] All these redefinitions of Indian water rights contradicted existing United States Supreme Court opinions, and the high court, by denying certiorari, allowed the Wyoming case to stand as good law.[18]

Ironically, Congress, normally a leader when it came to taking rights away from Native Americans, was appalled at the actions of the Supreme Court. It thereafter began backing agreements made by tribes with large users and offering federal financial support to develop large water projects. These resulted in the loss of some Indian water rights but the protection of others, and for the first time Native Americans shifted their legal focus from the courts to the federal legislature. Most of these comprehensive settlements occurred in the Southwest.[19]

Thus Indian efforts to exercise their rights to natural resources became a serious battle of significant legal proportions. Water, fishing, and hunting rights, so clearly defined in the past by treaties, statutes, and court opinions, began to unravel in the modern era. This was and continues to be a life-and-death struggle for Native American peoples, for without food and water, reservation and Indian life itself cannot be sustained.

Applying the New Tribalism to Indian Economics and Politics

The modern era of Indian rights provided the setting for raising a number of important legal issues that have yet to be resolved. Related to hunting, fishing, and water rights is the power of Native Americans to use the natural resources of the reservation. Exercising this right has taken tribes through a variety of legal venues, including scrutiny of leases, zoning, and taxation. The power to tax, as many tribal governments have discovered, is an important power, and the exercise of taxation has been controversial. Similarly, the ability of tribal governments to extend civil and criminal jusrisdiction to the reservation, regardless of who is involved, Indian or non-Indian, stretched the concepts of Indian self-determination as defined by the federal government. These economic and political issues were dramatically addressed in the past two decades.

During the late 1970s and the 1980s, Americans were surprised to learn that Indians own much of the nation's energy resources. West of the Mississippi River, 3 to 5 percent of oil and gas, 30 to 40 percent of coal, and 25 to 40 percent of the country's uranium reserves are on Indian reservations. The coal on Montana and North Dakota reservations alone contains fifteen times the energy resources of the Alaska North Slope's oil and gas fields. Although this looks like a bonanza, it is important to note that only approximately 40 out of over 300 federally recognized Native American nations own rights to energy resources. By the end of the 1980s, most of these resources were untapped, but private companies were anxious to move in and capture mineral rights on the reservations.[20]

Indian reaction to the sudden interest was divided. Some Native Americans opposed any mining developments, seeing their cultures being threatened. Others thought that mining would help provide jobs and funds for future economic development on the reservations. Many Indians were simply cynical. Said one Hopi, "Don't tell me about an energy crisis. I don't even have electricity in my village."[21]

What was important was that Indian leadership debated these issues. After they had gathered information, they quickly found many companies arranged grossly unfair long-term leases through the Department of the Interior, often without consulting with the tribes. When the tribes tried to void the leases, they were rebuffed at almost every turn. The result was a new legal thrust by Native American tribal governments into the energy field. First, they created their own organization, the Council of Energy Resource Tribes (CERT). Then Indian tribes began to monitor the exploitation of their own natural resources. Through the efforts of LaDonna Harris (Comanche), Allen Rowland (Northern Cheyenne), Earl Old Person (Blackfoot), Roland Johnson (Laguna Pueblo), Dale Vigil (Jicarilla Apache), Peter McDonald (Navajo), and others, CERT evolved into an international entity. For a time, CERT acted like an American OPEC, but as the 1980s progressed, CERT returned to more conventional methods of dealing with Indian resource issues.[22]

To combat lease problems, new federal laws addressed Indian concerns and included Indians in previously neglected natural-resource planning. In 1982, the Indian Mineral Development Act became law, giving the tribes the legal ability to produce natural resources from their reservations. Previously they could only receive royalties from leases. Now they could set up their own extractive companies.[23]

Just as important to the tribes as controlling what was being extracted from their reservations was having some say over what was being placed on or near their reservations. Also attractive to companies and states were reservation lands for the deposit of wastes, including nuclear wastes. In the same year that Congress passed the Indian Mineral Development Act, it created the Nuclear Waste Policy Act, which established regulatory procedures and standards.[24] Tribal governments petitioned to be classified as "affected." Once a group received this designation, it could participate in the decision making. The first tribe to be defined as "affected" was the Yakimas. They were then invited to review nuclear-waste-repository activities at the Hanford site near their Washington State reservation.

What the Yakimas found and other tribes subsequently discovered is that nuclear waste will not benefit their people. Moreover, the Department of Energy (DOE) was more concerned about putting the waste in the ground than deciding where the waste would be placed or determining its impact on Indian peoples. At Hanford, before 1982 and the passage of the Nuclear Waste Policy Act, DOE had tunnels dug into Gable Mountain without consulting the Yakima Nation or considering First Amendment ramifications. Gable Mountain is central to the Yakimas' religious beliefs, and the tribal council responded with a series of resolutions. One in particular was significant: The Yakimas banned the transportation of nuclear materials through their reservation. Eventually, the Yakimas were able to stop further desecration of Gable Mountain. Ironically, it is the states—in this case, the state of Washington—that now want to ally with Indian tribal governments to prevent nuclear wastes from being deposited on or near reservation lands.[25]

Another legal means to protect the Indian land base from exploitation involved the tribal governments in zoning. Tribal govnernments in the 1970s and 1980s enacted zoning ordinances covering entire reservations. This power, they argued, was sanctioned by *Williams v. Lee* (1959). The states were preempted from asserting power over reservations generally and over non-Indians residing on reservations specifically. Since 1959, the *Williams* precedent was eroded until it was effectively overruled in *Brendale v. Confederated Tribes and Bands of the Yakima Indian Nations* (1989).[26]

Previously, in *United States v. Mazurie* (1975), the Supreme Court had decided that tribal governments could regulate non-Indians only if they were specifically given that power by the federal government.[27] This was a substantiated departure from *Williams v. Lee*. In *Brendale*, the Yakima Indian Nation regulated the development of lands owned by Yakimas, other Indians, and non-Indians—all living on the Yakima Reservation. The Yakima Nation found itself in conflict with the Yakima County Planning Department, which disputed zoning on the reservation. The Supreme Court articulated a test allowing Indian zoning regulations to stand only if the ordinance protected the fundamental nature of the tribe. In a circuitous opinion that divided in several ways, the Court held that through zoning, the Yakimas could regulate areas of the reservation that were not under the control of non-Indians, but it could not zone mixed-use areas because they were already less related to Yakima culture.[28]

Modern tribalism to Indians meant reclaiming control over their lands. They tried to build coalitions and influence national energy policy through CERT. They used tribal ordinances to delay and actually force enactment of national legislation to guarantee Indian voices in nuclear waste disposal. Zoning regulations became another weapon to prevent unwelcome development. But perhaps the most important of the powers available to tribal governments was the power to tax—to tax companies extracting natural resources from Indian lands.

Indians and the Power to Tax

In the 1970s and 1980s, more and more tribal governments declared their resources to be the subject of plenary tribal authority, subject only to specific, congressionally mandated federal control. They wanted to manage their own reservation resources, but frequently, they ran into problems with long-term leases negotiated by the Department of the Interior. Not to be denied, tribal governments began to tax the mining companies. The general federal interest in reservations precluded state control, but little law developed regarding any kind of preemption of tribal taxing power. Moreover, the states were taxing mineral development. State taxes on non-Indians who extracted minerals from reservation lands were not preempted by general federal interests.[29]

The logical extension of the new tribalism into the economy of Indian nations and the lack of definitive legal standards moved Native Americans into the taxation arena. Three kinds of taxation were available to tribal governments. They could enact taxes on consumption, typically sales or use taxes. The most popular were tribal taxes on state tax–free cigarettes. Another kind of tax was the property tax, on housing, resource production, or real estate. A logical opportunity for Indian governments was severance taxes on mineral extractions and oil and gas production. The third kind of tax available was the income tax.[30]

Not known in 1970 was the extent to which tribal taxes preempted state taxes or state taxes preempted tribal taxes. Also unclear was whether the states could tax non-Indian activities on the reservations. Congress clearly had not addressed these issues, so it was left to the Supreme Court to decide them. The first case heard originated in an Arizona–Navajo dispute. Rosalind McClanahan, a Navajo who lived on the Navajo Reservation in Arizona, had $16.20 withheld from her paycheck by the state of Arizona, even though her income came completely from her work on the reservation. That is, the state of Arizona was attempting to tax the income that Navajos earned on their reservation. The Court held in *McClanahan v. State Tax Commission of Arizona* (1973) that no state could tax income that Indians earned on reservations because federal law preempted state action. Such a state tax interfered with tribal self-government and self-determination policies. Open to question was whether the state could tax non-Indians residing and deriving income on the reservation.[31]

In spite of the gray areas and the *McClanahan* decision, tribes were reluctant to implement tax ordinances. For example, the Navajos were not constituted as a taxing agent until 1974 when they created the Navajo Tax Commission. The Navajo Tribal Council then waited two more years before anyone was even appointed to the tax commission. Indian nations did not want to tax the income of their own people, mostly because Indian income was minimal. Similarly, collecting sales taxes and property taxes from their tribal members was impractical. The main reservation sources of income were severance taxes on large non-Indian corporations and consumption taxes on goods, usually cigarettes, that non-Indians bought because they were cheaper on reservations. Indian governments also feared restrictions, as they were concerned about the Indian Bill of Rights equal protection clause. Would tribes be forced to tax Indians and non-Indians alike?[32] After *Santa Clara Pueblo v. Martinez* (1978), that fear was reduced, but it remained nevertheless. Another concern was whether the state would be forced to tax the same products or items that Indians taxed, thereby driving away any income-producing development from the reservation. This latter fear was realized as early as 1976.

On both taxing fronts, the Supreme Court gave the states an entry onto reservations. In Montana, Joseph Wheeler, Jr., a Flathead who owned two discount cigarette stores on the Flathead Reservation, was arrested by state authorities for having sold cigarettes to non-Indians without the Montana state tax stamp affixed. In *Moe v. Confederated Salish & Kootenai Tribes of the Flathead Reservation* (1976), the Supreme Court allowed the states to tax Indian incomes on reservations that were derived from non-Indian commerce.[33]

Litigation also included severance taxes. The Jicarilla Apaches won a major victory in 1982 when their tax on oil and gas was upheld as an inherent aspect of their sovereignty and the federal government's policy of self-determination.[34] The state of New Mexico argued that only the state and municipalities could tax oil and gas production on reservations. When the Kerr-McGee Corporation challenged Navajo taxes on their mining activities, it lost. Put to rest was any notion that tribal governments could not tax non-Indians on reservations, but the Supreme Court left open the notion that if Indian taxes were "unfair," they could be struck down.[35]

These early limited tax victories for Native Americans, however, were substantially eroded within a decade. Borrowing from the *Moe* precedent, the state of New Mexico decided to tax non-Indian oil and gas producers on Indian reservations. In *Cotton Petroleum Corp. v. New Mexico* (1989),[36] the Supreme Court ruled that these kinds of state taxes were permissible, thereby threatening the very profitability of the product that the Indians sought to tax.

The only way out now for tribes is to own their own oil and gas and mineral companies or to negotiate with the states. Indian nations have begun moving in both these directions. As revenues dwindled in the late 1980s, tribal governments and the states were forced to reach accommodations with each other. The tax battlefields of the 1970s and 1980s are in a current state of truce, but hostilities are always near the surface.[37] Throughout these legal maneuvers, potential Fifth and Fourteenth Amendment federal protections for Indian tribes or the application of the Indian Bill of Rights to issues of taxation or just compensation failed to enter the dialogue.

Applying the New Tribalism to Indian Social Structures

Indian families in the 1970s continued to experience forced acculturation, but its proponents relied heavily on the legal system and social welfare agencies. Twenty-five to 35 percent of the children of Native American mothers and fathers throughout the United States were taken from them. In Idaho, out of every 1,000 children placed in foster care, 78 were Indian children, compared with 12 non-Indian children; in Maine the ratio was 76 to 4, and in Minnesota it was 58 to 4. In South Dakota, 40 percent of all adoptions involved Indian children being legally assigned to non-Indian adoptive parents. Seventeen percent of all Indian children in the United States lived at federal boarding schools away from their families.[38]

The taking of Indian children from their parents defied due process, but charges of physical abuse were almost never found against Native American parents. Violence toward children is not a part of Indian culture, regardless of the grinding poverty that many Native American families experience. Instead, social workers used social deprivation, emotional damage, or permissiveness as reasons for uprooting Indian children.[39] These were thinly veiled bureaucratic excuses for intolerance of a different culture. The welfare and Christian missionary bureaucracies refused to respect the extended family and tribal values of Native Americans. Indeed, congressional hearings singled out non-Indian Mormons in Utah, Idaho, and Arizona, as well as state social agencies in states with large Indian populations, with aggressively campaigning to take Indian children away from their parents, families, and tribal settings.

For example, in the early 1970s, a two-year-old Indian girl was taken away from her Indian mother by an Oregon state caseworker because the child was ''failing to thrive.'' True, the child showed no weight gain or height increase. The mother told the caseworker that her daughter threw up after each feeding. By the time it was learned that a doctor had prescribed the wrong formula for the child, Oregon's courts had taken the child away from her Indian mother. The caseworker refused to tell the mother where her child was. After protests by the Indian Studies Center in Portland, the state was forced to have doctors examine the child, and they found that the girl suffered from a hormonal deficiency. Finally, the Indian girl was reunited with her mother, and it was then discovered that the child had been physically abused by the non-Indian foster parents.[40]

Another example occurred on Pine Ridge Reservation in South Dakota. At Christmastime in 1972, two Wisconsin women missionaries asked the Sioux mother of three-year-old Benita Rowland if they could take her for a short visit to Wisconsin. Benita's mother signed what she thought was a permission slip, but it turned out to be an agreement to end her parental rights. When the child was not returned, Benita's mother protested. The Wisconsin Christian women offered to buy the child, and they sent this message: ''We have not taken Benita from you; you gave her physical birth, which we could not give, and we can give her opportunities which you could not give—so she belongs to both of us. But far more, she belongs to the Lord.''[41] During her childhood, Benita Rowland never was reunited with her mother.

Before 1978, the Indian parents and tribes of the Oregon two-year-old and the South Dakota three-year-old had no recourse against Indian child–stealing practices. All state courts subscribed to the ''best interests of the child'' doctrine if custody challenges made it into state courts, and the courts did not define Indian families as beyond immediate ken. Thus the Indian cultural view of the extended family was interpreted by American courts as detrimental. The Indian grandparent or aunt and uncle were treated as legal strangers with no direct interest in a contested child. Leaving a child with an aunt was construed by non-Indian judges to mean abandonment.[42]

Some states, however, recognized the Indian child welfare crisis for what it was and began to take appropriate steps. Michigan courts in 1973 articulated a landmark decision in Indian family law. Three Potawatomi children born of a Potawatomi father and a non-Indian mother were orphaned by the sudden murder-suicide of their parents, which occurred in Michigan, not on their Wisconsin reservation. The Michigan Department of Social Services took custody of the children and placed them in the home of a maternal non-Indian aunt and uncle. In accordance with tribal custom, the paternal great-uncle asked to adopt the children and was denied. A Michigan court attempted to resolve the dispute. In the process, it examined Potawatomi custom and discovered that Potawatomis live in kinship communities in which distant relatives are automatically responsible for child care and that adoption is not a legal concept because substitute parents are always available. The court held that the care of Indian children was a matter of tribal concern, an internal tribal affair within the exclusive jurisdiction of the tribe.[43]

Modern tribalism had at last attained a small victory in one state. It brought legal warriors to fight for Indian children. Many Indian parents and Native American communities were ill equipped to challenge the welfare decisions of the states and counties. These legal warriors helped Indian self-determination policies expand Indian sovereignty toward a strengthening of cultural integrity. Controversies over the custody of Indian children required a federal law or a Supreme Court determination on rights to cultural integrity or perhaps even cultural survival for Indian tribes and families.[44] Since the federal courts seemed less responsive than in the past and rarely did the Supreme Court consider family law a federal issue, Congress became the logical pressure point. The result, the Indian Child Welfare Act of 1978, was generally satisfying for many Native Americans.

The Indian Child Welfare Act of 1978 (ICWA) culminated four years of congressional hearings.[45] Signed into law by President Jimmy Carter, the ICWA contains three sections. Its purpose is to "protect the best interests of Indian children and to promote the stability and security of Indian tribes and families by the establishment of minimum Federal standards for the removal of Indian children from their families."[46]

Title I, the bulwark of the act, outlines procedures for preventing private and public agencies from disrupting Indian families. Child-custody proceedings involving Indian children on reservations are to be within the exclusive jurisdiction of Indian courts. To take an Indian child away from its parents requires proving "serious emotional or physical damage to the child," a higher standard than that found in most jurisdictions.[47] The ICWA also provides for strict preferences for placement if a child is orphaned or its emotional or physical well-being is in jeopardy. Most important, the tribe is guaranteed notice of any proceeding involving one of its own children.[48]

Several nuances in the Indian Child Welfare Act indicate challenges to existing laws and rights. Indian parents, child custodians, or the child itself is guaranteed the right to court-appointed counsel, a new right not provided in the Indian Bill of Rights that heretofore was not guaranteed to individual Indians in Indian court proceedings. Perhaps even more revolutionary are the provisions for the retrocession of Public Law 280 for Indian tribes involved in child-custody proceedings. The act also encourages state and private welfare bureaucracies to make it their duty to prevent the destruction of Indian families. Keeping records of custody proceedings involving Indian children is required in all states so that the results of the act can be monitored.[49]

The Indian Child Welfare Act constituted a beginning of hope for many Native American families and tribes. At the least, it has forced child agencies and state courts to look at their actions more closely. After only a decade of the ICWA's operation, it may be too soon to tell if it has accomplished its goals, but there are signs that not every institution is in compliance and that the act contains loopholes. The biggest problem may be identifying the child as an Indian, particularly if the child welfare agency does not investigate fully or if the state court does not pay attention to the requirements of the Indian Child Welfare Act.[50] In addition, tribal courts and tribal governments must pass family law codes, set up new family law courts, and train personnel. Nevertheless, this path-breaking federal legislation continues to provide a means toward realizing modern tribalism for Native Americans.

The American Indian Religious Freedom Act of 1978

The year 1978 also provided the setting for another federal attempt to protect Indian culture that was much less successful than the Indian

Child Welfare Act. Congress decided to consider the impact of various governmental policies on Native American religious practices. Testimony before a select committee of Congress chaired by Senator James Abourezk of South Dakota elicited numerous comments concerning the denial of Indian religious freedom throughout the United States.

On one occasion before the committee, Stephen Rios, a member of the Juaneño Band of California Mission Indians, explained a problem for his people in Nevada and California: They are not allowed access to their most sacred shrine—a hot springs—from which they believe they emerged onto this earth. They could not visit it because it is located on a United States Navy base, the Channel Lake Naval Weapons Center, where geothermal development is taking place right on the religious site. Rios told how his tribe had attempted to preserve the sacred site and to gain access to it. The navy simply would not respond, so the Juaneño leader took the issue to his state legislator, who reluctantly offered an excuse for his inaction. "Well, Steve," he said, "I don't even understand my own religion. I don't see how you can expect me to understand Indian religion, as strange and as difficult as it is."[51]

At the hearings, the testimony of Crow tribal leader Barney Old Coyote offered perhaps the most succinct explanation as to why Indian religions differ from Christian religions and why non-Indians have trouble understanding Indian religions: "The area of worship cannot be delineated from social, political, culture, and other areas of Indian lifestyle, including his general outlook upon economic and resource development." Worship is, according to Old Coyote, "an integral part of the Indian way of life and culture which cannot be separated from the whole. This oneness of Indian life seems to be the basic difference between the Indian and non-Indians of a dominant society."[52] To Native Americans, their traditional religions are synonymous with their entire being, their environment and culture. American law, judges, and legislators have had a difficult time grasping this concept, particularly during the 1970s and 1980s.

Congress found serious violations for American Indians of what are traditionally thought to be First Amendment rights. Native Americans were being denied access to religious sites; they were not being allowed to use and possess sacred objects; and their freedom to worship through traditional means was restricted. To help solve this problem, Congress passed, and President Carter signed, the American Indian Religious Freedom Act of 1978 (AIRFA).[53]

The act contains only two provisions. The first defines the proper role of the United States government: "Henceforth it shall be the policy of the United States to protect and preserve for American Indians their inherent right of freedom to believe, express, and exercise the traditional religions." This section applies to access to "sacred sites required in their religions, including cemeteries," the use of sacred objects in religious ceremonies, and the practice of traditional Indian rites.[54]

The second provision requires presidential action. The president must order federal agencies to evaluate their laws in consultation with traditional native religious leaders "in order to determine appropriate changes necessary to protect and preserve Native American religious cultural rights and practices." After one year, the president is required to issue a report to Congress documenting any policy changes.[55]

What does this crucial First Amendment reiteration include? Clearly, it is a restatement for Native Americans of the rights to believe, express, and exercise their traditional religions. These rights specifically include access to sacred sites, such as Channel Lake for California Mission Indians; use and possession of sacred objects, such as eagle feathers and peyote; and freedom to worship in traditional ceremonies, such as in sweat lodges in prisons. AIRFA also issued an important directive to federal agencies: They must consult with actual practicing Indian religious leaders, rather than BIA bureaucrats or anthropologists.

Indians welcomed this new law and immediately began negotiations with the federal government and other institutions regarding sacred objects. Dialogues started throughout the United States. For example, the Zunis were successful in working with the Fish and Wildlife Service to obtain protected bird feathers and animals. Religious artifacts stolen from the tribe were recovered with the help of a United States attorney in New York City, and agreements were reached with a number of museums over the proper display of sacred objects.[56] Other tribes, however, met resistance, and AIRFA was not able to be of much help to them.

The act does have deficiencies. The legislation substantively treats doctrine and ceremony, but only those beliefs of the past actually practiced and continually renewed are to be preserved. In regard to ceremony, "access" is the crucial word. Visitation rights to sites are ensured, but this represents mere immediacies—in other words, short-term practices. The law does not preserve "reality"; only "personalty," such as a sacred object, is protected. It does not place an umbrella around holistic philosophy; it recognizes only limited doctrine because religion is not considered a dynamic, evolving aspect of culture. Religious changes and new doctrine are thus denied. Perhaps the act's greatest weakness is that it has no "teeth." That is, persons restricting or denying Indians their free exercise of their religion cannot be punished or sued.[57]

Even before the American Indian Religious Freedom Act was passed, native peoples had challenged non-Indian activities that inhibited religious practices. Court confrontations usually concerned efforts by tribes to preserve sacred sites from economic development by non-Indians. In *Sequoyah v. Tennessee Valley Authority* (1980), the Cherokees unsuccessfully sought to prevent construction of the Tellico Dam on the Little Tennessee River. They argued that their traditional hunting grounds and areas used to gather medicinal and religious objects would be destroyed, but the appellate court ruled that the dam could be built because the valley was not "central" to Cherokee religious practices. AIRFA was barely mentioned.[58] In *Badoni v. Higginson* (1980), Navajos tried to force the lowering of the water held in the Glen Canyon Dam on the Colorado River in southern Utah and to restrict tourist activity at the Rainbow Bridge National Monument. But the Navajos lost because the court considered their religious interests less important than the need to use the dam for electric power. The court also ruled that the Navajos' access to the Rainbow Bridge National Monument was not prevented, and so tourism did not have to be restricted. Again AIRFA was not controlling because it was passed during the evolution of the suit.[59]

The Supreme Court finally got a chance to consider the legal impact of the American Indian Religious Freedom Act in *Lyng v. Northwest Indian Cemetery Protective Ass'n* (1988).[60] For more than eleven years, the Yurok, Karok, and Tolowa Indian nations had been trying to stop the Forest Service from paving a road and harvesting timber on the Hoopa Valley Indian Reservation and in the Six Rivers National Forest in northern California. The Forest Service commissioned a report that recommended preventing the completion of the road because it would open to the general public an area central to private Indian religious practices, but the Forest Service went ahead with its plans anyway. In the federal district court, the judge ruled that the Forest Service should not take action because it would establish a burden on the Indians' free exercise of their religion, and the appellate court affirmed the district court's decision.

However, the Supreme Court overruled everyone and gave the Forest Service permission to pave and harvest. In a 5 to 3 opinion written by Justice Sandra Day O'Connor, the American Indian Religious Freedom Act was eviscerated. The Court created a new two-pronged test: Indians must prove that they were coerced into violating their beliefs or penalized for practicing their beliefs. If neither could be proved, then no religious infringement had occurred. This is not the same free-exercise standard articulated for any other persons or groups in the United States.

The impact of *Lyng* is significant. Of immediate importance is that the Forest Service began destroying a region essential to 5,000 Indians for their religious practices. Access was no longer a concept that prevented the destruction of sacred sites. Interference with Indian religion could occur whether or not the government's action was important. Restrictions on dams might mean the loss of electricity for thousands, but a road into a secluded section of a forest had a minimal impact. Most important, the Court once again decided that the First Amendment did not apply to Native Americans, and furthermore, AIRFA offered no substitute statutory guarantee.[61] If sacred sites could not be saved, then sacred objects, religious rites, and even cultural practices could not be protected.

By 1990, the protections hinted at by AIRFA and the basic rights in the First Amendment were of minimal use to Native Americans, and the highest court finally made the ultimate first strike, throwing into question the legality of the Native American Church and all Indian traditional religious practices. Two Indians living in Portland, Oregon, Alfred Smith and Galen Black, were fired from their jobs with a private drug rehabilitation organization because they had used peyote during a ceremony of their Native American Church. They applied for unemployment compensation, but they were denied benefits because the state of Oregon determined that they had been discharged for misconduct. After a great deal of litigation, the Supreme Court heard the case and found against Smith and Black and all of Oregon's state courts.[62] In so doing, the Court struck a major blow against First Amendment religious freedoms for Native Americans and opened the door for state agencies to begin another assault on the Native American Church.

Five judges, led by Justice Antonin Scalia, voted to accept the notion that the use of peyote could be prohibited under a state criminal statute, even though Oregon had not enforced this law. According to the majority, the Oregon statute did not infringe on individual rights supposedly guaranteed by the free-exercise provisions of the First Amendment or AIRFA for practitioners of the Native American Church. No persons could hold their religious practices against the state's right to regulate the use of drugs. Peyote is recognized as a sacramental element in the Native American Church, but that did not matter here. Indeed, the Court seemed uninformed about the nature of peyote. Adding further discriminatory language to the holding, the majority in dicta noted that the physical act of drinking wine in a Christian ceremony, presumably even by minors in violation of state alcohol statutes, would be acceptable.[63]

All the previous tests used for assessing the limits of the free-exercise clause and other Indian religion cases were summarily disposed. Court members appeared unhappy with one another as the justices penned cutting words in their opinions suggesting that their colleagues did not know what they were doing. It remains to be seen whether *Employment Division, Department of Human Resources of Oregon et al. v. Alfred L. Smith et al.* (1990) will be the controlling statement on Native American religious freedoms. If it is, there will be no Indian religious freedom in the United States.

■　　■　　■

Peace and harmony for Native Americans in search of their legal rights were not available during the 1980s and the first years of the 1990s. Indian rights are in retreat. Hunting, fishing, and water rights were restricted. Abilities to obtain control over reservation resources, whether through taxation or zoning or extensions of tribal jurisdictions, were contested with each step. Of particular concern were efforts in the 1980s to inhibit Native American social and religious life. Even with Congress attempting to control these powerful movements, such as through the Indian Child Welfare Act and the American Indian Religious Freedom Act, there were few causes for celebration among Indians seeking to exercise fully what most Americans would see as traditional rights by every person.

By 1991 and the bicentennial of the federal Bill of Rights, Native Americans are poised to move on the offensive. The grounds were shifting once again. Native Americans want desperately to improve their economic lot, and a federal gaming act poses questions, worries, and opportunities. With the beginnings of a new civil rights movement, Indian demands for recognition of their rights have come in a form that most non-Indians can understand. The legal issue is the repatriation of Indian skeletal remains and burial goods from institutions—universities, historical societies, and even the most revered of museums in the United States, the Smithsonian Institution. Non-Indians can identify with Indians in their grief and their resolve, and what Native Americans want has empathetic substantive and symbolic characteristics. Already the lines have been drawn in what promises to be a renewal of the struggle by native peoples for their full rights within the limited sovereignty of their collective existence.

10

Epilogue

The federal Bill of Rights celebrated its bicentennial in 1991. Much was made of this event by non-Indians, and many myths were told.[1] Native Americans did not generally participate in the celebration; instead, they assessed their legal situation and acted accordingly.

Two hundred years of legal disappointments and legal disasters for native peoples are the legacy of the bicentennial of the federal Bill of Rights. Indians continue to be the only persons living in the United States not covered by Bill of Rights provisions, and the Indian Bill of Rights is subject to manipulation and distortion by Congress and the federal courts. Rights are seen by Indians as their collective responsibility. In her novel about the legal struggles of Sioux rancher John Tatekeya, Elizabeth Cook-Lynn explains this legal meaning. At a trial brought about by Tatekeya's willingness to prosecute his white neighbors for cattle rustling, Jason Big Pipe testifies against him. Although Tatekeya wins, he is disturbed. As Cook-Lynn observes,

> The ugly testimony given by Jason at the trial had successively shifted the guilt of the white man's theft to the people, *Oyate,* to themselves and each other and all who clung tenaciously to the notion that, even amidst devastating change, the Dakotahs were people who were obliged to be responsible to one another.[2]

In 1991, as in 1791, the legal struggles continued for Native Americans. Issues of sovereignty abound. Economic security, political hegemony, and cultural respect are central to any understanding of Indian attempts to retain and advance group and individual rights in their own societies or in the majority society. Of particular importance in 1991 were the federal gaming statute and legal challenges waged by many Indian nations to obtain the skeletal remains and burial goods of their ancestors.

Rights and Resolve: Activating the Law for Native American Interests, 1991 and Beyond

It is difficult to project the historical landscape of the present into the future. For the United States of the 1980s, a carefree ''get-it-while-you-can'' approach to American life seemed comfortable while it lasted, but by 1991 the postponed economic ramifications were already beginning to be felt. Even the war in the Persian Gulf in 1991 did not prove to be a positive lasting experience for the American psyche.

Indians found little solace during the 1980s; those rights regained in the 1930s and 1970s were restricted once again. Conditions on reservations seemed perilously close to turn-of-the-century hardships. One difference, however, is that Native Americans were able to withstand attacks on their rights and sometimes even win. Tribal governments and lawyers keep up the pressure to restrain rights preemptions and regain rights previously lost. New legal issues and new Indian leadership are emerging in the 1990s, but unlike the past, Indians now are prepared to fight for their rights.

Indian Law Leadership

In order to survive within the laws and legal system of non-Indians, Native Americans successfully developed over the past decade a new cadre of law leaders who are directing the attempts by Indians to make sure their rights are not lost. There are many, but two who stand out are Wilma Mankiller, principal chief of the Cherokee Nation of Oklahoma, and John Echo-Hawk, Pawnee attorney and director of the Native American Rights Fund. Together they represent the political and legal merger necessary in the struggle to retain Indian sovereignty and Indian rights.

The Cherokee Nation of Oklahoma, whose capital, Tahlequah, is located in the ''green country'' of northeastern Oklahoma, is the third largest Indian nation, after those of the Navajos and Lumbees. It has a population of more than 120,000, an annual budget of $54 million, and 800 employees. In 1983, Ross Swimmer ran for principal chief and Wilma Mankiller for deputy chief in tribal government elections. Two years after their election, Swimmer was chosen to be the Commissioner of Indian Affairs, and Mankiller assumed the principal chieftainship. Mankiller then ran successfully for election as principal chief in 1987 and was reelected with 83 percent of the vote in June 1991, to a second four-year term.

Wilma Mankiller is one of eleven children who grew up first on her father's allotted Cherokee lands in Oklahoma. But in 1957, when Mankiller was eleven, she and her family became victims of the Indian relocation program. She found herself in high-crime urban poverty in San Francisco. The Alcatraz protest and the treatment of Native Americans by the United States persuaded Mankiller to return to her people and to work with them to improve their lives and reclaim their rights. She went back to Oklahoma in 1977 to put her training and energy into Cherokee community development, obtaining grants for housing improvements, health clinics, and extensions of water lines to Cherokee homes. Politics then called.

To Mankiller, the realization and practical application of what her Cherokee elders had taught her, that "we are all interdependent," guides her leadership. She looks ahead to the future and sees politics and law working together. Says Peterson Zah, former director of Navajo Legal Services (Dinebeiina Nahiilna Be Agiditahe [DNA]) and twice elected the chairman of the Navajo Nation, "Wilma is a breath of fresh air in Indian leadership. She is a visionary who is very aggressive about achieving the goals she has in mind for her people."

Mankiller herself is confident. "We can look back over the 500 years since Columbus stumbled onto this continent and see utter devastation among our people. But as we approach the 21st century," she states, "we are very hopeful." Mankiller points out that "despite everything, we survive in 1991 as a culturally distinct group. Our tribal institutions are strong. And I think we can be confident that, 500 years from now, someone like Wilma Mankiller will say that our languages and ceremonies from time immemorial still survive."[3]

Law leadership in the 1980s and 1990s is provided by the many Native American attorneys recently trained in several American law schools. Indian law programs begun in the 1970s hit their stride in the 1980s and include, most notably, those at the University of New Mexico; the University of Oklahoma, home of the *American Indian Law Review;* the University of Colorado; and the University of Iowa. In 1960, there were only 40 Indian lawyers, but by 1991 there were at least 700. Making headlines and significant Indian law contributions are associated with the Native American Rights Fund, which is based in Boulder, Colorado, and directed by Pawnee attorney John Echo-Hawk.

The Native American Rights Fund, or NARF, was founded in 1970 from a grant by the Ford Foundation. In its first decade, NARF filed 1,900 lawsuits, and it has continued to be active in the courts. These challenges for Indian rights and those of Native Hawaiians have ranged from significant land cases to prisoner religious rights, school regulations concerning hair length, and the repatriation of Indian skeletal remains. Of special interest to NARF are disputes involving sovereignty over reservations, including Indian rights to tax, zone, police, and develop their own lands. Water rights and the leasing of mines are important as well. According to Echo-Hawk, NARF is conducting a successful legal strategy, not unlike the NAACP's efforts for African-American rights. "For us," says Echo-Hawk, "it is a matter of survival. We hope these court decisions will settle once and for all the question of Indian rights. We are seeking no more than what is rightfully ours, spelled out in the law."[4]

The law in 1991 is being used in the fight for Indian rights. John Echo-Hawk and Wilma Mankiller each represent an element of the new law leadership that is attempting to exercise Native American rights collectively, rights that should have naturally evolved over time from interpretations of the federal Bill of Rights and the Indian Bill of Rights that have yet to be achieved. Many legal issues are currently facing Native Americans, several of which are rooted in the past but never seem to go away. In order to apply twenty-first century legal methods to these problems, Indians have turned to reclamation: a reclamation of their history, their culture, and their rights, through the myriad legal avenues available. Two contemporary legal flash points involve gaming and repatriation.

Gaming

One of the two most important contemporary legal issues facing Native Americans concerns legalized gambling. Indians, of course, have been gambling for hundreds of years. Indians like a good game of chance, and this is a part of their cultural heritage. Whether it is the Arikara awl game or the Pawnee Ghost Dance hand game, Native Americans consider gaming a part of life.

The goal of federal self-determination policies is to help Native Americans identify economic opportunities, and the federal government has focused on Indian economic development both on and off reservations. This is regarded by acculturationists as another possible attempt at legalized termination, but it also happens to coincide with the desire of those Indians who want to reclaim their sovereignty, exercise Native American rights, and provide for a better life for Indian people.

Gambling in the United States cuts across many political spectrums. It defies categorization. But there is no doubt that non-Indians want to gamble and that they will pay significant amounts of money to do so. Nevada has long held the monopoly on gambling, and American legal tradition has left the regulation of this "social sin" primarily to the states. Thus when economic hard times struck several states in the 1970s and 1980s, a number of them, such as New Jersey and South Dakota, liberalized their gambling laws. Many states already allowed some forms of gambling, and so they simply added to this tradition.

Indians found themselves with an excellent economic opportunity, and some tribes cashed in. Bingo halls sprang up around the United States. Although states tried to prevent them from operating, they were unsuccessful. Several federal courts ruled that the states could not tamper with gambling sponsored by Indian tribes, and although there were few general federal laws on the subject, there were enough to exempt the area from state regulation.[5]

Finally, in 1987, the Supreme Court decided to hear a case arising out of attempts by California—a Public Law 280 state—to prevent two Indian tribes, the Cabazon and Morongo bands of Mission Indians, from sponsoring gambling in Riverside County. The Cabazon Band operated a club where card games were played for prizes, and both tribes managed bingo halls. Riverside County wanted to apply its ordinances to tribal gambling activities, and the state intervened to apply its criminal statutes. But federal district and appellate courts supported the tribes.

By 1987, gambling was not an insignificant economic activity. In the previous year, the Cabazons, a tribe of twenty-five members, had grossed $5 million from their Indio, California, bingo hall. Bingo halls operated by Seminoles in four Florida locations brought in $45 million;

Cherokees made $10 million at their North Carolina facility; and at Jamestown in northern California, a small tribe of only six Rancheria (Mission) Indians took in $9 million at their Chicken Ranch bingo hall.[6] In all, bingo games supplied up to 90 percent of Indian gambling revenues.

In *Cabazon Band of Mission Indians et al. v. California* (1988), the Supreme Court granted constitutional approval to Indian commercial gambling.[7] In a 6–3 opinion, Justice Byron White ruled that Public Law 280 and the Organized Crime Control Act of 1970 were not expressions of congressional intent to allow the states to regulate Indian gambling. Since there was no express congressional word and there were no exceptional circumstances that required state or county gambling laws to be applied to Indian bingo operations, and there was a clear federal policy to encourage economic development for Native Americans, the Cabazon Band was free to operate its business as it saw fit.

Cabazon reestablished the line of case law begun with *McClanahan v. State Tax Commission of Arizona* (1973) and briefly reversed by *Moe v. Confederated Salish & Kootenai Tribes of the Flathead Reservation* (1976). *McClanahan* held that state laws did not apply to Indians on reservations unless expressly stated by federal law, and *Moe* allowed state taxation of cigarette purchases made by non-Indians on reservations. Justice White saw gambling as generating value, an almost manufacturing kind of analogy, whereas he viewed smoke shops as being in business primarily to avoid state taxation. Although *Cabazon* could be read as a victory for tribal sovereignty, there are plenty of warnings in the opinion that the standards invoked could be reversed by a different set of facts.[8]

Because of the rising revenues and the concerns of many states, Congress responded by enacting the comprehensive Indian Gaming Regulatory Act (IGRA) the year after the *Cabazon* decision.[9] The act defines three kinds of gambling. Class I gaming includes those traditional forms of Indian gambling that are usually a part of tribal celebrations and social games for minimal value prizes. They are not subject to regulation except by tribal governments. Class II covers bingo and card games for prizes, and Class III gaming basically encompasses casino gambling, including baccarat, blackjack, and slot machines. To regulate Class II and III gaming, the National Indian Gaming Commission was established in the Department of the Interior.[10]

There are many requirements for tribes that wish to set up bingo halls and casinos, and they are specific even to the point of requiring the tribes to take no less than 30 percent and no more than 40 percent of the gross revenues. In some ways, IGRA can be read as insulting to tribes, since it treats them as if they were children. At the end of three years, if a tribe has not had any problems, it can regulate its Class II gaming activities on its own. The commission is authorized to issue a "certificate of self-regulation."[11]

Of particular importance is that section of the IGRA outlining the role of the states and Class III gaming. If a state authorizes this kind of gambling, even if it is for charities, Indians can set up casinos. In this situation, the states and tribes must negotiate a tribal–state compact. The act makes it clear that the state cannot avoid negotiating and that the tribe does have the right to set up casinos.

Casinos can be placed on any Indian land held by a tribe that has been granted trust protection by the Secretary of the Interior. However, there are stipulations that limit this action. The Secretary of the Interior can refuse to allow a casino to be built on Indian land if he determines that gaming would be detrimental to the surrounding communities and if the governor of the state disapproves. The secretary must also sanction any tribal–state compacts. Clearly, self-determination is a concept that flows both in and out of this document.[12]

The Indian Gaming Regulatory Act has yet to be tested in court, but all the signs point to this future development. Already controversy surrounds the act and the role of the Secretary of the Interior. Only two states—Hawaii and Utah—bar all types of gambling and are immune from the act. Some, like Connecticut, have tried to resist both Class II and Class III gaming, but they have been unsuccessful.[13]

The controversy most likely to test the act is developing in Iowa. In the 1980s, Iowa significantly liberalized its gambling laws to allow for horse- and dog-race betting and riverboat casino-style gambling. The Santee Sioux tribe of Nebraska bought an option on land in Council Bluffs, Iowa, near Interstate Highways 29 and 80 and the large metropolitan area of Omaha. In conjunction with Harvey's, a major casino operation in Nevada, the Santees then asked Secretary of the Interior Manual Lujan for permission to build a $67.5 million casino, convention center, and hotel that would employ 1,000 people. Governor Terry Branstad of Iowa and Congressman Peter Hoagland of Nebraska opposed this plan, as did several communities in Nebraska. The city of

Council Bluffs and Pottawattamie County, however, embraced the Santees' casino idea. The Santees believed that they had met the criteria imposed by IGRA by obtaining immediate community support, and they tried to force Iowa to negotiate a tribal–state compact.

At the last minute, the Mesquakie Indian Settlement in central Iowa was persuaded to oppose the Santee plan. The Mesquakies operate a bingo hall and were considering whether to expand it into a casino. (Located near Tama, Iowa, the Mesquakie Settlement is not near a large metropolitan area.) Then Louis LaRose, tribal chairman of the Nebraska-based Winnebagos, who own a small amount of land on the eastern side of the Missouri River, wrote a letter opposing the Santee plan. He stated that the Santee casino would detract from a proposed Winnebago casino to be built on isolated Winnebago land near Sloan, Iowa. Other Winnebago tribal leaders immediately disavowed the letter. The only other tribe with trust lands in Iowa, the Omahas, who also planned to build a casino on their Iowa land near Onawa, refused to oppose the Santee.

On December 19, 1991, the Department of the Interior announced that Secretary Lujan had rejected the Santee plan. Although the governor of Iowa objected to the idea of a tribe from another state buying land in Iowa to cash in on Iowa's gambling laws, the secretary said he did not base his decision on that notion. That is, he could not very well object to buying land for a casino off a reservation, since he had already approved one such purchase by the Forest County Potawatomis of northern Wisconsin, who bought land for a bingo parlor in downtown Milwaukee. Rather, Lujan stated that he was protecting the interests of tribes already in Iowa from outside tribes. Thus the secretary broadly construed the language of the Indian Gaming Regulatory Act to define "community" as encompassing both the surrounding towns and cities as well as the state's Indian communities. Santee leaders are now considering a lawsuit and also are trying to work out a partnership with the Omahas.[14]

Gambling has been an economic windfall for many tribes. By 1991, Indian country gambling had grossed over $400 million a year. Some tribes, such as the 103-member Shakopee Mdewakanton Sioux, who happen to own land near Minneapolis and St. Paul, Minnesota, are doing very well. All members receive a monthly check for about $2,000 and a variety of financial benefits, including a new home if they lack one, full college scholarships, and trust funds for their children that may be as high as $80,000 by the time the child reaches age eighteen. For others, gambling has brought problems. The bingo hall of the Yakima Indian Nation filed for bankruptcy in 1991 in part because of an embezzlement scandal. Some Indians believe that outside developers are taking advantage of the situation. "The new-found self-determination and authority being granted to the tribes," argues Vernon Bellecourt, a leader of AIM and a Minnesota Chippewa, "in many cases is being used and abused against our own people."[15] Nevertheless, the right to gamble on Indian land remains a right available to all Americans; at least it was in 1991.

Repatriation

Indian leaders and lawyers have enlightened the general public about America's own killing fields. During the twentieth century, massive numbers of Indian skeletons were dug up, stored, and frequently put on display by museums, universities, tourist traps, and government agencies. Estimates are that the remains of as many as 2 million Native Americans have been disturbed.[16]

The stories of these misdeeds are grizzly. The Army Medical Museum kept more than 4,000 Indian heads taken from burial grounds, scaffolds, and fresh graves. Government agents decapitated Cheyennes and Arapahoes who were murdered at Sand Creek and Modoc leaders who were executed in Oregon. Northern Cheyenne leaders discovered over 18,000 Indian bodies stored by the Smithsonian Institution. A shocking revelation about the American Museum of Natural History told of the museum's duplicity in arranging a fake funeral for a dead Inuit so that his son would not find out that the institution had stolen his father's body. As Franz Boas, one of the United States' most famous anthropologists, noted, "It is most unpleasant work to steal bones from graves, but what is the use, someone has to do it."[17]

In addition to the hoards of skeletons kept by the institutions for research infrequently conducted, display cases, and entertainment purposes, Congress learned that many unrelated funerary objects, sacred objects, and the cultural patrimony of tribes were taken from them and deposited in museums and federal agencies, even though every state has statutes regulating cemeteries, protecting graves from desecration, and making burglary, robbery, and larceny a crime. Several outrageous state court decisions tried to define Indian remains separately from all other skeletal remains. In California, a state court ruled that an Indian cemetery was not a "cemetery" for state purposes. In Ohio, pre-European settlement remains were defined as not being "human" remains under Ohio's grave-robbing statute.[18] A loophole in state laws that left unmarked graves unprotected allowed some to claim that grave robbing was legal, but in the past decade, thirty-four states approved new statutes to extend coverage to unmarked graves.[19]

Extending coverage to unmarked graves did little to solve the problems of tribes wanting *repatriation,* the return of their dead, funerary objects, sacred objects, and cultural patrimony. Fifth and First Amendment considerations seemed inoperative given past and recent Supreme Court opinions. What was needed, reasoned Indian legalists, were new state and federal statutes that forced the return of skeletal remains, all tribal objects, and patrimony.

Since 1989, four states—Arizona, Hawaii, Nebraska, and Kansas—have passed repatriation laws. The most comprehensive act concerning the return of skeletal remains and funerary objects was adopted by Nebraska, and although the Nebraska State Historical Society fought passage of the law and its implementation, Pawnees were eventually successful in reburying over 400 of their ancestors. Arizona's 1990 statute covers human remains as well as sacred objects and tribal patrimony.[20]

These acts, however, may be preempted by a federal law passed and signed in 1990, the Native American Graves Protection and Repatriation Act (NAGPRA).[21] Daniel Inouye, senator from Hawaii and chair of the Senate Select Committee on Indian Affairs, introduced the legislation by explaining that it was designed to address flagrant violations of the rights of America's first peoples. He noted that "when human remains are displayed in museums or historical societies, it is never the bones of white soldiers or the first European settlers that came to this continent that are lying in glass cases. It is Indian remains." To Senator Inouye, "The message that this sends to the world is that Indians are culturally and physically different from and inferior to non-Indians. This is racism."[22]

The Native American Graves Protection and Repatriation Act is comprehensive, stipulating that within five years all museums, universities, and agencies (except for the Smithsonian Institution, which is covered under separate, less stringent legislation) must make inventories of human remains and associated funerary objects in cooperation with tribal governments and Indian religious leaders. Also, within three years, these institutions must compile a summary of all holdings of unassociated funerary objects, sacred objects, and cultural patrimony. Tribes are to have access to these materials and may request them and be granted repatriation.

A Repatriation Review Committee also is stipulated by NAGPRA, which will consist of seven members, including Native American religious leaders and representatives from museums and scientific organizations. The review committee is to monitor the inventory and summary processes, act as an arbitrator in disputes between interested parties, and make its own inventory of culturally unidentifiable human remains. In addition, it is to make recommendation of culturally unidentifiable remains.

Unlike the American Indian Religious Freedom Act, the Native American Graves Protection and Repatriation Act mandates significant penalties for those institutions that ignore its provisions. Fines can be assessed against museums and universities based on the value of the item concerned, damages, and the number of violations. A tribe or an individual Indian may take institutions to federal court for failure to abide by NAGPRA. Future archaeological excavation and patrimony acquisitions also are under new rules requiring permission and notice.[23]

Given the storm of controversy caused by Native Americans' insistence on repatriation, archaeologists, amateur pothunters, historical societies, and museum personnel will likely challenge the constitutionality of the Native American Graves Protection and Repatriation Act. Because NAGPRA deals directly with Fifth and First Amendment issues that have a history of judicial scrutiny, the outcome cannot be predictable. On the side of those who oppose repatriation is the tradition that federal courts have been reluctant to grant Native Americans coverage under the Fifth and First Amendments. Indians can find solace in knowing that the current Supreme Court is generally deferential to congressional intent, and Congress's views are clearly stated in the Native American Graves Protection and Repatriation Act. Like gaming and the economic opportunities that gambling may or may not represent, repatriation brings to the law a contemporary challenge for Native Americans to achieve a cultural healing heretofore denied their nations.

■   ■   ■

In 1935 Black Wolf, a Northern Cheyenne elder of Lame Deer, Montana, agreed to share his knowledge of the past with legal anthropologist E. Adamson Hoebel and jurisprudential scholar Karl N. Llewellyn. Black Wolf related a story about one Cheyenne man, Sticks Everything Under His Belt, who held his individual rights against the world and his people.[24]

Around 1866, Sticks Everything Under His Belt decided to hunt buffalo alone. He said that he was hunting for himself; he did not believe that tribal tradition and Cheyenne rules applied to him. In essence, he was declaring himself to be above or beyond his own people. The soldier chiefs and the tribal chiefs met to decide how to handle this matter of an individual's declaring a "right" against the nation, and they concluded that Sticks Everything Under His Belt should be ostracized. Anyone who wanted to help him would have to give a sun dance.

For two years, Sticks Everything Under His Belt lived alone in the camp. People refused to recognize him. Finally, his brother-in-law, a chief, took pity on his relative. He asked his wife to cook a feast, and he invited all the chiefs to their tipi. When everyone had arrived, the brother-in-law announced that he believed it was time to reconsider Sticks Everything Under His Belt's penalty. "I am going to give a Sun Dance to bring him back in. I beg you to let him come back to the tribe, for he has suffered long enough. This Sun Dance will be a great one."[25] The chiefs were pleased and agreed. They set the conditions. "Ha-ho, ha-ho [Thank you, thank you]. . . . Let him remember that he will be bound by whatever rules the soldiers [the society of Cheyenne military leaders] lay down for the tribe. He may not say he is outside of them. He has been out of the tribe for a long time. If he remembers these things, he may come back."[26]

Sticks Everything Under His Belt came to the lodge. He smoked with all the chiefs, and then he fainted. When he recovered, he stated, "From now on I am going to run with the tribe. Everything the people say, I shall stay right by it. My brother-in-law has done a great thing. He is going to punish himself in the Sun Dance to bring me back. He won't do it alone, for I am going in, too."[27]

Sticks Everything Under His Belt decided that his desire to exercise individual rights had been foolish and that his loyalty to his nation was more important. The legal system of the Northern Cheyennes considered group entitlements and a consensus on the Cheyenne way so central that it bestowed one of its most extreme punishments on Sticks Everything Under His Belt. And yet atonement was achieved, and the wayward Cheyenne rejoined the nation without malice.

More than 120 years later, new visitors journeyed to Lame Deer, Montana. This time the law of the Cheyenne was explored not by legal scholars but by Hand Made Films, a movie company. David Seals's novel *Powwow Highway* was made into an award-winning film through the path-breaking efforts of Indian writers, filmmakers, and actors.[28]

Again, as in Black Wolf's recollections, two themes were merged. In the film *Powwow Highway,* Buddy Red Bow tries to save his tribe from an unscrupulous mining company represented by another Indian, Sandy Youngblood, while Red Bow's childhood friend Philbert Bono, a big, happy dreamer, decides to make a vision quest to find himself and his nation. The mining company conspires with the BIA and local law-enforcement authorities to frame Buddy's sister Bonnie Red Bow in a drug case. Ten years earlier, Bonnie left the reservation and her people's traditions to go to Sante Fe. There she is jailed on the drug charge and so calls her brother.

Even though he has not heard from her in years (he does not even know that she has two children), Buddy takes off for Sante Fe with Philbert. Buddy was supposed to deliver $2,000 of the tribal government's money to buy bulls for the tribe's ranch, but instead he impulsively takes the money in order to pay for the trip to Santa Fe.

Why do they go? Because, as Philbert says, "We are Cheyenne." Why is Philbert searching for the sacred mountain, Sweet Butte, which he calls the most sacred place in America? "Nobody cares about history these days, but I do." In their effort to preserve the land of the Northern Cheyenne, to give reverence to the sacredness of traditional Cheyenne life, to fight pollution, and to counteract violence, Buddy and Philbert make their way to Santa Fe, get Bonnie out of jail, reunite her with her children and her heritage, and find the money necessary to buy the bulls. They are confronted with numerous perils, a corrupt system of non-Cheyenne laws, and other Indians who have given up. As Buddy observes, "Often the problems never change, nor do The People."[29]

Individual Indian rights are expressed in *Powwow Highway,* and they are positively portrayed because they help save the nation, express collective entitlements and group consensus, and renew the Cheyenne way. Buddy Red Bow, like Sticks Everything Under His Belt, achieves renewal with his people, and Philbert Bono (his new name is Whirlwind

Dreamer), like the Cheyenne brother-in-law over 100 years earlier, sacrifices for the well-being and healing of the nation.

For these Northern Cheyennes, and Indian people generally, individual rights may or may not be the answer to what is legally important to their world. An alien legal code that defined those rights in the nineteenth century was not helpful to Native Americans, and the promise of law reforms in the twentieth century too often ended in failure, failures that the federal Bill of Rights did not prevent.

A bill of rights represents many things to different people. In the United States, the federal Bill of Rights is regarded by many as a crucial element in the nation's foundation of freedom and liberty. Individuals and groups may hold against the world basic, fundamental rights, or they may use it as a check against the abuse of power by government. Liberty and freedom require the balancing of coercive power against individual and collective rights.

But for the native peoples of the United States, the federal Bill of Rights does not provide protection from the powers of government, be it federal, state, local, or even tribal. Indians have been denied the individual rights and group rights given to all other inhabitants of the United States. Sovereignty and all it entails have not been deemed compatible with the Bill of Rights by the non-Indian majority that dominates the American legal system. Only Congress can prevent full-scale denials to Native Americans of those basic human rights laid down in the federal Bill of Rights.

The history of Indians and the Bill of Rights is a story of continuity and change. From the earlier rights held by the first peoples of America to the Old and New Colonialisms they experienced in the nineteenth century, from the legal and human survival of Native Americans in the early twentieth century to the introduction of constitutionalism to Indian nations, from the dualities of Indian law and the federal cancer of termination present in the legal system to the creation of the Indian Bill of Rights, and from the evolution of modern tribalism and Indian rights in retreat to contemporary legal issues, it is a continuing history that must be constantly told and retold.

Notes

Chapter 1

1. Burten D. Fretz, "The Bill of Rights and American Indian Tribal Governments," *Natural Resources Journal* 6 (October 1966): 613–616.

2. For a brief survey of the historical uniqueness of Indian law, see Wilcomb E. Washburn, "The Historical Context of American Indian Legal Problems," *Law and Contemporary Problems* 40 (Winter 1976): 12–24.

3. Ted C. Lewellen, *Political Anthropology: An Introduction* (Westport, Conn.: Bergin and Garvey, 1983); June Starr and Jane F. Collier, eds., *History and Power in the Study of Law: New Directions in Legal Anthropology* (Ithaca, N.Y.: Cornell University Press, 1989).

4. Lewellen, *Political Anthropology*, pp. 85–90. See also the introductory sections of Allen C. Turner, "Evolution, Assimilation, and State Control of Gambling in Indian Country: Is *Cabazon v. California* an Assimilationist Wolf in Preemption Clothing?" *Idaho Law Review* 24, no. 2 (1987–1988): 318–326.

5. E. Adamson Hoebel, *The Law of Primitive Man: A Study in Comparative Legal Dynamics* (Cambridge, Mass.: Harvard University Press, 1954); E. Adamson Hoebel and Karl N. Llewellyn, *The Cheyenne Way: Conflict and Case Law in Primitive Jurisprudence* (Norman: University of Oklahoma Press, 1941); and E. Adamson Hoebel and Ernest Wallace, *The Comanches: Lords of the South Plains* (Norman: University of Oklahoma Press, 1952), are three of his significant works. See also Hoebel's outrage at Indian termination policy in "To End Their Status," in *The Indian in Modern America,* ed. David A. Baerreis (Madison: State Historical Society of Wisconsin, 1956), pp. 1–15.

Although Hoebel subscribed to a number of outmoded concepts from early anthropological theory, including the ranking of cultures from the "primitive" or "savage" to the "complex" or "civilized," these weaknesses must be noted and necessarily forgiven. His contributions to identifying indigenous concepts of rights are too important to dismiss and are not dependent on flawed racial classifications.

6. Hoebel, *Law of Primitive Man,* pp. 4–17.

7. Ibid.; Robert H. Lowie, "Property Rights and Coercive Powers of Plains Indian Military Society," *Journal of Legal and Political Sociology* 1 (April 1943): 59–71; Donald L. Burnett, Jr., "An Historical Analysis of the 1968 'Indian Civil Rights' Act," *Harvard Journal of Legislation* 9 (May 1972): 577–578.

8. Hoebel, *Law of Primitive Man,* pp. 18, 20–28.

9. Ibid., pp. 48–50.

10. John Phillip Reid, *A Law of Blood: The Primitive Law of the Chreokee Nation* (New York: New York University Press, 1970), p. 38.

11. Hoebel, *Law of Primitive Man,* pp. 51–55.

12. Anthony F. C. Wallace, *The Death and Rebirth of the Seneca* (New York: Vintage Books, 1972), p. 25; Hoebel and Llewellyn, *Cheyenne Way,* pp. 233–235, 279; Reid, *Law of Blood,* p. 40; Robert A. Fairbanks, "The Cheyenne and Their Law: A Positivist Inquiry," *Arkansas Law Review* 32 (Fall 1978): 403–445; Robert A. Fairbanks, "A Discussion of the Nation-State Status of American Indian Tribes: A Case Study of the Cheyenne Nation," *American Indian Journal* 3 (October 1977): 2–24.

13. I am indebted to Richard White for suggesting this term to me during my quandary as to how to describe and distinguish individual rights from groups rights in post-1791 Indian society.

14. Vine Deloria, Jr., "Self-Determination and the Concept of Sovereignty," in *Economic Development in American Indian Reservations,* ed. Roxanne Dunbar Ortiz (Albuquerque: University of New Mexico Native American Studies, 1979), p. 22.

15. Roxanne Dunbar Ortiz, "Sources of Underdevelopment," in Ortiz, ed., *American Indian Reservations,* p. 61.

16. John Howard Clinebell and Jim Thomson, "Sovereignty and Self-Determination: The Rights of Native Americans Under International Law," *Buffalo Law Review* 27 (Fall 1978): 670–682.

17. Stephen Cornell, *The Return of the Native: American Indian Political Resurgence* (New York: Oxford University Press, 1988), pp. 45–50. See also Gail H. Landsman and Laurence M. Hauptman, "Commentary on *Sovereignty and Symbol: Indian–White Conflict at Ganienkeh," Ethnohistory* 38 (Summer 1991): 304–310.

18. Judith Jarvis Thomson, *The Realm of Rights* (Cambridge, Mass.: Harvard University Press, 1990), pp. 2, 348.

19. Russel Lawrence Barsh and James Youngblood Henderson, *The Road: Indian Tribes and Political Liberty* (Berkeley: University of California Press,

1980), pp. 270–282; E. J. Hobsbawm, *Nations and Nationalism Since 1780* (Cambridge: Cambridge University Press, 1991).

20. Charles F. Wilkinson, *American Indians, Time, and the Law: Native Societies in a Modern Constitutional Democracy* (New Haven, Conn.: Yale University Press, 1987), pp. 54–55.

21. Vine Deloria, Jr., and Clifford M. Lytle, *The Nations Within: The Past and Future of American Indian Sovereignty* (New York: Pantheon Books, 1984), p. 2.

22. Arrell Morgan Gibson, *The American Indian: Prehistory to the Present* (Lexington, Mass.: Heath, 1980), pp. 16–35; Peter Farb, *Man's Rise to Civilization: The Cultural Ascent of the Indians of North America* (New York: Dutton, 1968), pp. 3–15.

23. Farb, *Man's Rise to Civilization*, pp. 16–154; Gibson, *American Indian*, pp. 37–89; Angie Debo, *A History of the Indians of the United States* (Norman: University of Oklahoma Press, 1970), pp. 3–18. See also Dean Snow, *The American Indians: Their Archaeology and Prehistory* (London: Thames and Hudson, 1976); Edward H. Spicer, *A Short History of the Indians of the United States* (New York: Van Nostrand Reinhold, 1969); D'Arcy McNickle, *They Came Here First: The Epic of the American Indian* (Philadelphia: Lippincott, 1949), esp. pp. 75–91.

24. See generally Robert A. Williams, Jr., *The American Indian in Western Legal Thought: The Discourses of Conquest* (New York: Oxford University Press, 1990); L. C. Green and Olive P. Dickason, *The Law of Nations and the New World* (Edmonton: University of Alberta Press, 1989), esp. pp. 144–185; Wilcomb E. Washburn, *Red Man's Land/White Man's Law: A Study of the Past and Present Status of the American Indian* (New York: Scribner, 1971).

25. *Barkham's Case* 1 Va. Company of London 71–87 (1621). See also "Introduction," in *Records of the Virginia Company of London*, ed. S. M. Kingsbury (Washington, D.C.: Library of Congress, 1933), vol. 1; Williams, *American Indian in Western Legal Thought*, pp. 205–218, esp. n.120.

26. Gibson, *American Indian*, pp. 91–216; Henry F. Dobyns, "Indians in the Colonial Spanish Borderlands," in *Indians in American History*, ed. Frederick E. Hoxie (Arlington Heights, Ill.: Harlan-Davidson, 1988), pp. 95–116; James Axtell, "Colonial America Without the Indians: Counterfactual Reflections," in *The American Indian: Past and Present*, 4th ed., ed. Roger L. Nichols (New York: McGraw-Hill, 1992), pp. 1–13.

27. Williams, *American Indian in Western Legal Thought*, pp. 287–323.

28. Simon J. Ortiz, "The Land and the People Are Speaking (Sthehgaadzih Tzikannah Tehneh eh Yaatrah Naitra Guh [Once again, Value and Respect Must Come About])," in Ortiz, ed., *American Indian Reservations*, p. 11.

29. Ibid., p. 12.

Chapter 2

1. Ben Kindle's Winter Count for the year 1791, from Martha Warren Beckwith, "Mythology of the Oglala Sioux," *Journal of American Folklore* 43 (1930): 399–442, cited in Russel Lawrence Barsh and James Youngblood Henderson, *The Road: Indian Tribes and Political Liberty* (Berkeley: University of California Press, 1980), p. 31.

2. Dragging Canoe, Treaty of Fort Stanwix Councils of 1768, in *The Portable North American Indian Reader,* ed. Frederick W. Turner III (New York: Viking, 1974), p. 244.

3. U.S., *Statutes at Large* 7 (September 17, 1778): 13–15.

4. Donald A. Grinde, Jr., *The Iroquois and the Founding of the American Nation* (San Francisco: Indian Historian Press, 1977); Bruce E. Johansen, *Forgotten Founders: Benjamin Franklin, the Iroquois, and the Rationale for the American Revolution* (Ipswich, Mass.: Gambit, 1982).

5. U.S. Constitution, art. I, sec. 2, para. 3.

6. Ibid., art. I, sec. 8, para. 3.

7. Ibid., art. I, sec. 10, para. 1 and 3.

8. Ibid., art. VI, para. 2.

9. U.S., *Statutes at Large* 1 (July 22, 1790): 137–138.

10. See U.S., *Statutes at Large* 1 (March 1, 1793): 329–332; U.S., *Statutes at Large* 1 (May 19, 1796): 469–474; U.S., *Statutes at Large* 1 (March 3, 1799): 743–749.

11. U.S., *Statutes at Large* 2 (January 17, 1800): 6–7.

12. U.S., *Statutes at Large* 2 (May 13, 1800): 85. For a discussion of dependency theory and how it applies to United States relationships with three tribes, see Richard White, *Roots of Dependency: Subsistence, Environment, and Social Change Among the Choctaws, Pawnees, and Navajos* (Lincoln: University of Nebraska Press, 1983).

13. U.S., *Statutes at Large* 3 (March 3, 1817): 383.

14. U.S., *Statutes at Large* 3 (March 3, 1819): 516–517.

15. U.S., *Statutes at Large* 4 (June 30, 1834): 729–735.

16. U.S., *Statutes at Large* 4 (May 28, 1830): 411–412.

17. Ibid.

18. Ibid.

19. John R. Wunder, "No More Treaties: The Resolution of 1871 and the Alteration of Indian Rights to Their Homelands," in *Working the Range: Essays on the History of Western Land Management and the Environment,* ed. John R. Wunder (Westport, Conn.: Greenwood Press, 1985), p. 41.

20. Quoted in William T. Hagan, *American Indians,* rev. ed. (Chicago: University of Chicago Press, 1979), pp. 70–71.

21. *Johnson v. M'Intosh* 21 U.S. 543 (1823). Charles F. Wilkinson identified

Johnson v. M'Intosh along with the two Cherokee cases as the Marshall trilogy constituting the most important historical developments in Native American–United States relationships (*American Indians, Time, and the Law: Native Societies in a Modern Constitutional Democracy* [New Haven, Conn.: Yale University Press, 1987], pp. 23–24).

22. *Cherokee Nation v. State of Georgia* 30 U.S. 1 (1831).

23. *Worcester v. Georgia* 31 U.S. 515 (1832).

24. Ibid.

25. Wilkinson, *American Indians,* pp. 23–24; Charles F. Wilkinson, "The Place of Indian Law in Constitutional Law and History" (manuscript), pp. 107–108, 126. The other three cases cited more often than *Worcester v. Georgia* are *Marbury v. Madison* 5 U.S. 127 (1803), *McCulloch v. Maryland* 17 U.S. 316 (1819), and *United States v. Perez* 22 U.S. 579 (1824).

26. Wilkinson, "Place of Indian Law," pp. 103–129; Joseph C. Burke, "The Cherokee Cases: A Study in Law, Politics, and Morality," *Stanford Law Review* 21 (February 1969): 500–531.

27. *The Kansas Indians* 72 U.S. 737 (1866).

28. U.S., *Statutes at Large* 16 (March 3, 1871): 544–571 at 566.

29. Wilcomb Washburn, *The Indian in America* (New York: Harper & Row, 1975), pp. 97–98; Francis Paul Prucha, *The Great Father: The United States Government and the American Indians,* 2 vols. (Lincoln: University of Nebraska Press, 1984), vol. 1, pp. 527–533; Wunder, "No More Treaties," pp. 39–56.

30. Hagan, *American Indians,* pp. 108–110. For a cogent analysis of Sioux geopolitics before the signing of the Treaty of Fort Laramie, see Richard White, "The Winning of the West: The Eighteenth and Nineteenth Centuries," *Journal of American History* 65 (September 1978): 319–343. Se also "Fort Laramie Treaty of 1868," in *The Great Sioux Nation: Sitting in Judgment on America,* ed. Roxanne Dunbar Ortiz (Berkeley, Calif.: Moon Books, 1977), pp. 94–99; Peter John Powell, "The Sacred Treaty," in Ortiz, ed., *Great Sioux Nation,* pp. 105–109.

31. U.S., *Statutes at Large* 15 (February 16, 1869): 635–647. See also K. Kirke Kickingbird, ed., *Treaties & Agreements and the Proceedings of the Treaties and Agreements of the Tribes and Bands of the Sioux Nation* (Washington, D.C.: Institute for the Development of Indian Law, 1974), pp. 74–100.

32. Quoted in Doane Robinson, *A History of the Dakota or Sioux Indians* (Minneapolis: Ross and Haines, 1956), p. 397.

33. *Turner v. American Baptist Missionary Union* 24 F. 346 (1852).

34. *Congressional Globe,* 41st Cong., 3rd sess., 1871, p. 733. See also Wunder, "No More Treaties," pp. 47–48.

35. U.S., *Statutes at Large* 24 (February 8, 1887): 388–391 [Dawes Severalty Act or General Allotment Act].

36. Ibid.; Rennard Strickland and Charles F. Wilkinson, eds., *Felix S. Co-*

hen's Handbook of Federal Indian Law (Charlottesville, Va.: Michie Bobbs-Merrill, 1982), pp. 78–79.

37. Hagan, *American Indians,* pp. 141–148.

38. Robert M. Kvasnicka and Herman J. Viola, eds., *The Commissioners of Indian Affairs, 1824–1977* (Lincoln: University of Nebraska Press, 1979).

39. U.S., *Statutes at Large* 18 (June 22, 1874): 566–571.

40. Ibid., Title XXVIII Indians, chap. 2, pp. 364–369 at 365 (sec. 2088).

41. U.S., Commissioner of Indian Affairs, *Report,* 1872, p. 6.

42. U.S., Commissioner of Indian Affairs, *Report,* 1890, pp. viii, lvii.

43. U.S. Congress, House of Representatives, "Policy and Administration of Indian Affairs," in *Report on Indians Taxed and Not Taxed, at the Eleventh Census, 1890,* misc. doc. no. 340, 52nd Cong., 1st sess., 1894, pt. 15, p. 68, quoted in Strickland and Wilkinson, eds., *Cohen's Indian Law,* p. 177.

44. *Dobbs v. United States* 33 Ct. Cl. 317 (1898).

45. *Tully v. United States* 32 Ct. Cl. 13 (1896).

46. Hagan, *American Indians,* pp. 137–140.

47. Strickland and Wilkinson, eds., *Cohen's Indian Law,* p. 141.

48. *Ex parte Crow Dog* 109 U.S. 556 (1883).

49. Bernard Floyd Hyatt, "A Legal Legacy for Statehood: The Development of the Territorial Judicial System in Dakota Territory, 1861–1889," 2 vols. (Ph.D. diss., Texas Tech University, 1987), vol. 2, pp. 552–556.

50. *Ex parte Crow Dog* 109 U.S. 556 (1883). See also Sidney L. Harring, "Crow Dog's Case: A Chapter in the Legal History of Tribal Sovereignty," *American Indian Law Review* 14 (November 2, 1989): 191–239.

51. U.S., *Statutes at Large* 23 (March 3, 1885): 362–385 at 385 [Major Crimes Act].

52. *United States v. Wheeler* 435 U.S. 313 (1978).

53. *Talton v. Mayes* 163 U.S. 376 (1896).

54. Hagan, *American Indians,* pp. 124–127.

55. *Quick Bear v. Leupp* 210 U.S. 50 (1908).

56. *Missouri, Kansas and Texas Railway Co. v. Roberts* 152 U.S. 117–118 (1894), quoted in Vine Deloria, Jr., "Within and Without: The American Indian and the Constitution" (manuscript), p. 43.

57. U.S., *Statutes at Large* 15 (August 25, 1868): 581–587 [Treaty of Medicine Lodge Creek of 1867].

58. For a specific discussion of *Lone Wolf v. Hitchcock,* see Ann Laquer Estin, "*Lone Wolf v. Hitchcock:* The Long Shadow," in *The Aggressions of Civilization: Federal Indian Policy Since the 1880s,* ed. Sandra L. Cadwalader and Vine Deloria, Jr. (Philadelphia: Temple University Press, 1984), pp. 215–245.

59. U.S., Treaty of Medicine Lodge Creek of 1867, art. XII, p. 585.

60. Estin, "*Lone Wolf v. Hitchcock,*" pp. 224–226.

61. *Lone Wolf v. Hitchcock* 187 U.S. 553 (1903).

62. N. Scott Momaday, *The Way to Rainy Mountain* (Albuquerque: University of New Mexico Press, 1969), p. 44.

63. Ibid. Also symbolic of this era is *Standing Bear v. Crook* 25 F. 695 (1879), in which the Northern Ponca chief had to prove he was not an Indian to stay in his homelands.

Chapter 3

1. *United States v. Joseph* 94 U.S. 616 (1876).

2. *United States v. Lucero* 1 N.M. Terr. 444–445 (1869).

3. *United States v. Sandoval* 231 U.S. 28 (1913). See also David H. Getches and Charles F. Wilkinson, *Federal Indian Law: Cases and Materials,* 2nd ed. (St. Paul: West, 1986), pp. 197–200; W. Richard West, Jr., and Kevin Gover, "The Struggle for Indian Civil Rights," in *Indians in American History,* ed. Frederick E. Hoxie (Arlington Heights, Ill.: Harlan-Davidson, 1988), pp. 277–278.

4. U.S., *Statutes at Large* 31 (May 31, 1900): 221–248 at 246.

5. U.S., *Statutes at Large* 31 (March 3, 1901): 1058–1085 at 1058, 1083–1084.

6. U.S., *Statutes at Large* 32 (May 27, 1902): 245–277 at 275.

7. Angie Debo, *And Still the Waters Run* (Princeton, N.J.: Princeton University Press, 1940). On her struggle to unravel and put into print the revelations of corruption in Oklahoma's Indian land dispersal, see *Indians, Outlaws, and Angie Debo,* PBS video, prod. WGBH, WNET, and KCET, 1989.

8. U.S., *Statutes at Large* 34 (May 8, 1906): 182–183 [Burke Act of 1906].

9. *Elk v. Wilkins* 112 U.S. 94 (1884). See also Michael T. Smith, "The History of Indian Citizenship," *Great Plains Journal* 10 (Fall 1970): 25–35.

10. U.S., *Statutes at Large* 34 (June 21, 1906): 325–384.

11. U.S., *Statutes at Large* 34 (March 1, 1907): 1015–1052.

12. U.S., *Statutes at Large* 34 (March 2, 1907): 1221–1222.

13. U.S., *Statutes at Large* 35 (May 29, 1908): 444–458.

14. U.S., *Statutes at Large* 35 (March 3, 1909): 781–815 at 783.

15. U.S., *Statutes at Large* 36 (April 4, 1910): 269–289 at 271.

16. U.S., *Statutes at Large* 37 (February 14, 1913): 678–679.

17. U.S., *Statutes at Large* 38 (August 1, 1914): 582–608 at 583–587.

18. U.S., *Statutes at Large* 39 (May 18, 1916): 123–159 at 128.

19. U.S., *Statutes at Large* 40 (May 25, 1918): 561–592 at 564–565.

20. U.S., *Statutes at Large* 41 (June 30, 1919): 3–34 at 3–9. See also William T. Hagan, *American Indians,* rev. ed. (Chicago: University of Chicago Press, 1979), pp. 148–149.

21. Quoted in Hagan, *American Indians,* p. 147.

22. U.S., *Statutes at Large* 41 (February 14, 1920): 408–434 at 410–411.

23. *Piper v. Big Pine School District* 193 Cal. 664 (1924).

24. *Grant v. Michaels* 94 Mont. 452 (1933); *Crawford v. School District No. 7* 68 Or. 388 (1913).

25. U.S., *Statutes at Large* 41 (March 3, 1921): 1225–1249 at 1225–1227, 1231–1232.

26. U.S., *Statutes at Large* 43 (May 29, 1924): 244.

27. U.S., *Statutes at Large* 43 (June 2, 1924): 253.

28. Frederick E. Hoxie, *A Final Promise: The Campaign to Assimilate the Indians, 1880–1920* (Lincoln: University of Nebraska Press, 1984), p. 175.

29. U.S., Commissioner of Indian Affairs, *Report,* 1912, pp. 5–6, quoted in Hoxie, *Final Promise.*

30. See, for example, Arizona in *Porter v. Hall* 34 Ariz. 308 (1928) or Minnesota in *Opsahl v. Johnson* 138 Minn. 42 (1917). See also Vine Deloria, Jr., and Clifford M. Lytle, *American Indians, American Justice* (Austin: University of Texas Press, 1983), pp. 222–226.

31. *Winton v. Amos* 255 U.S. 373 (1921).

32. *Winters v. United States* 207 U.S. 564 (1908).

33. *United States v. Winans* 198 U.S. 371 (1905).

34. Ibid.

35. *Winters v. United States* 207 U.S. 564 (1908).

36. Ibid.

37. Hoxie, *Final Promise,* pp. 184–185.

38. Lawrence C. Kelly, ''Charles Henry Burke, 1921–1929,'' in *The Commissioners of Indian Affairs, 1824–1977,* ed. Robert M. Kvasnicka and Herman J. Viola (Lincoln: University of Nebraska Press, 1979), pp. 251–261.

39. U.S. Department of the Interior, Office of Indian Affairs, Commissioner Charles H. Burke to agency superintendents, Circular No. 1665, April 26, 1921, quoted in Monroe E. Price, *Law and the American Indian: Readings, Notes, and Cases* (Indianapolis: Bobbs-Merrill, 1973), pp. 700–701.

40. U.S., *Statutes at Large* 43 (June 7, 1924): 636–642 [Pueblo Lands Act].

41. Vine Deloria, Jr., ''The Indian Rights Association: An Appraisal,'' in *The Aggressions of Civilization: Federal Indian Policy Since the 1880s,* ed. Sandra L. Cadwalader and Vine Deloria, Jr. (Philadelphia: Temple University Press, 1984), p. 15.

42. Lewis Meriam and Henry Roe Cloud, *The Problem of Indian Administration* (Baltimore: Johns Hopkins University Press, 1928) [hereafter referred to as Meriam Report].

43. Francis Paul Prucha, *The Great Father: The United States Government and American Indians,* 2 vols. (Lincoln: University of Nebraska Press, 1984), vol. 2, p. 809.

44. Meriam Report, pp. 3, 9, 282, 287; Lawrence M. Hauptman, "The Indian Reorganization Act," in Cadwalader and Deloria, eds., *Aggressions of Civilization,* p. 135; Hoxie, *Final Promise,* p. 242.

45. Meriam Report, pp. 8, 11, 15; *Felix S. Cohen's Handbook of Federal Indian Law* (Albuquerque: University of New Mexico Press, 1942), p. 240; Hoxie, *Final Promise,* p. 242; West and Gover, "Indian Civil Rights," p. 280.

46. Meriam Report, pp. 3, 14, 16–17, 40–41; Hauptman, "Indian Reorganization Act," p. 134; West and Gover, "Indian Civil Rights," p. 280; Hoxie, *Final Promise,* p. 242.

47. Prucha, *Great Father,* vol. 2, pp. 810–811, 1018, 1082; Meriam Report, p. 15; Vine Deloria, Jr., "'Congress in Its Wisdom': The Course of Indian Legislation," in Cadwalader and Deloria, eds., *Aggressions of Civilization,* p. 113; Frederick E. Hoxie, "The Curious Story of Reformers and the American Indians," in Hoxie, ed., *Indians in American History,* p. 221.

48. Deloria and Lytle, *American Indians,* p. 12.

49. James E. Officer, "The Indian Service and Its Evolution," in Cadwalader and Deloria, eds., *Aggressions of Civilization,* p. 70.

50. Lucy Kramer Cohen, Charlotte Lloyd Walkup, and Benjamin Reifel, "Felix Cohen and the Adoption of the IRA," in *Indian Self-Rule: First-Hand Accounts of Indian–White Relations from Roosevelt to Reagan,* ed. Kenneth R. Philp (Salt Lake City: Howe Brothers, 1986), p. 73.

51. Alvin M. Josephy, Jr., "Modern America and the Indian," in Howie, ed., *Indians in American History,* p. 253; West and Gover, "Indian Civil Rights," p. 280.

52. Rennard Strickland and Charles F. Wilkinson, eds., *Felix S. Cohen's Handbook of Federal Indian Law* (Charlottesville, Va.: Michie Bobbs-Merrill, 1982), p. 26.

53. Wilcomb E. Washburn, *Red Man's Land/White Man's Law: A Study of the Past and Present Status of the American Indian* (New York: Scribner, 1971), p. 77.

54. Prucha, *Great Father,* vol. 2, pp. 811–812; Officer, "Indian Service," p. 70; John Painter, Robert L. Bennett, E. Reeseman Fryer, and Graham Holmes, "Implementing the IRA," in Philp, ed., *Indian Self-Rule,* p. 80.

55. James Welch, "Surviving," in *The Portable North American Indian Reader,* ed. Frederick W. Turner III (New York: Viking, 1974), p. 597 [emphasis added].

Chapter 4

1. Lawrence C. Kelly, "John Collier and the Indian New Deal: An Assessment," in *Indian–White Relations: A Persistent Paradox,* ed. Jane F. Smith and

Robert M. Kvasnicka (Washington, D.C.: Howard University Press, 1976), pp. 231–232.

2. Rudolph X. Foley, "The Origins of the Indian Reorganization Act of 1934" (Ph.D. diss., Fordham University, 1937), pp. 101–102, 205–230, 250–251.

3. Ibid., pp. 39, 42.

4. Roger Bromert, "The Sioux and the Indian-CCC," *South Dakota History* 8 (Fall 1978): 341, 343; Roger Bromert, "The Sioux and the Indian New Deal, 1933–1946" (Ph.D. diss., University of Toledo, 1980), p. 87.

5. Tully Hunter, "The Catawba of South Carolina, 1930–1962: The Failure of Twentieth Century American Indian Policy" (M.A. thesis, Clemson University, 1986), p. 57; Foley, "Origins of the IRA," pp. 45, 47.

6. Foley, "Origins of the IRA," pp. 67, 94.

7. Oliver La Farge, "The American Indian's Revenge," *Current History* 40 (May 1934): 163.

8. Peter M. Wright, "John Collier and the Oklahoma Indian Welfare Act of 1936," *Chronicles of Oklahoma* 50 (Autumn 1972): 352.

9. Michael T. Smith, "The Wheeler–Howard Act of 1934: The Indian New Deal," *Journal of the West* 10 (July 1971): 521; La Farge, "American Indian's Revenge," p. 164; D'Arcy McNickle, "The Indian New Deal as Mirror of the Future," in *Political Organization of Native North Americans,* ed. Ernest L. Schusky (Washington, D.C.: University Press of America, 1980), p. 108; William Hughes, "Indians on a New Trail," *Catholic World* 139 (July 1934): 464.

10. Smith, "Wheeler–Howard Act," p. 524.

11. U.S., *Statutes at Large* 47 (June 28, 1932): 336–337.

12. D'Arcy McNickle, "The American Indian Today," *Missouri Archaeologist* 5 (September 1939): 2.

13. Foley, "Origins of the IRA," pp. 17, 20–21.

14. Kenneth R. Philp, *John Collier's Crusade for Indian Reform, 1920–1954* (Tucson: University of Arizona Press, 1977), pp. 131–132; Elizabeth Green, "Indian Minorities Under the American New Deal," *Pacific Affairs* 8 (December 1935): 420; Jay B. Nash, ed., *The New Day for the Indians: A Survey of the Working of the Indian Reorganization Act of 1934* (New York: Academy Press, 1938), p. 26.

15. Quoted in McNickle, "Indian New Deal," p. 109.

16. Philp, *Collier's Crusade,* p. 132.

17. U.S., *Statutes at Large* 48 (May 21, 1934): 787. See also Philp, *Collier's Crusade,* p. 133; Nash, ed., *New Day for the Indians,* pp. 26–27; McNickle, "Indian New Deal," p. 109.

18. Nash, ed., *New Day for the Indians,* p. 27.

19. Philp, *Collier's Crusade,* pp. 122–126.

20. John Collier, "The Genesis and Philosophy of the Indian Reorganization Act," in *Indian Affairs and the Indian Reorganization Act: The Twenty Year Record,* ed. William H. Kelly (Tucson: University of Arizona, 1954), p. 7.

21. U.S. Congress, House of Representatives, H.R. 7902, 1934, Records of the Indian Reorganization Act, Records Group 75, National Archives, Washington, D.C.; Kenneth R. Philp, "John Collier and the Controversy over the Wheeler–Howard Bill," in Smith and Kvasnicka, eds., *Indian–White Relations,* pp. 178–184; Kelly, "Collier and the Indian New Deal," pp. 233–234.

22. W. David Baird, "Commentary," in Smith and Kvasnicka, eds., *Indian–White Relations,* pp. 215–221; Philp, "Collier and the Wheeler–Howard Bill," p. 175; Philp, *Collier's Crusade,* pp. 170–171; Wright, "Collier and the Oklahoma Indian Welfare Act," p. 356.

23. Frank Ducheneaux, "The Indian Reorganization Act and the Cheyenne River Sioux," *American Indian Journal* 2 (August 1976): 9. See also "The American Indian—Tribal Sovereignty and Civil Rights," *Iowa Law Review* 51 (Spring 1966): 654–659; Lynn Kickingbird, "Attitudes Toward the Indian Reorganization Bill," *American Indian Journal* 2 (July 1976): 9, 11–12; Foley, "Origins of the IRA," pp. 68–69.

24. Smith, "Wheeler–Howard Act," p. 527; Bromert, "Sioux and the Indian New Deal," pp. 58–59.

25. Joseph R. Garry, "The Indian Reorganization Act and the Withdrawal Program," in Kelly, ed., *Indian Affairs,* p. 35.

26. U.S., *Statutes at Large* 48 (June 18, 1934): 984–988 [Indian Reorganization Act or Wheeler–Howard Act].

27. U.S., *Statutes at Large* 49 (June 26, 1936): 1967–1968 [Oklahoma Indian Welfare Act]; U.S., *Statutes at Large* 49 (May 1, 1936): 1250–1251 [Alaska Indian Reorganization Act].

28. Lawrence C. Kelly, "The Indian Reorganization Act: The Dream and the Reality," *Pacific Historical Review* 44 (August 1975): 303.

29. U.S., Indian Reorganization Act, pp. 576–582.

30. Philp, *Collier's Crusade,* p. 118; Heinrich Krieger, "Principles of the Indian Law and the Act of June 18, 1934," *George Washington Law Review* 3 (March 1935): 296–299; Kelly, "Indian Reorganization Act," p. 297; Kelly, "Collier and the Indian New Deal," p. 235; Curtis Berkey, "The Legislative History of the Indian Reorganization Act," *American Indian Journal* 2 (July 1976): 15–22; McNickle, "Indian New Deal," pp. 110–111.

31. Kelly, "Indian Reorganization Act," pp. 301–303.

32. U.S., *Statutes at Large* 49 (June 15, 1935): 378.

33. Theodore H. Haas, *Ten Years of Tribal Government Under I.R.A.,* Tribal Relations Pamphlets, no. 1 (Chicago: United States Indian Service, 1947), p. 15. There is confusion regarding the votes and interpretations by historians. This same election is reported by Kenneth R. Philp as forty-three no votes, nine

yes votes, and sixty-two voting absences on the Santa Ysabel Reservation (*Collier's Crusade*, p. 162).

34. Kelly, "Indian Reorganization Act," pp. 304–305.

35. Philp, *Collier's Crusade*, pp. 181–183; Haas, *Tribal Government Under I.R.A.*, p. 28; Kelly, "Indian Reorganization Act," p. 305.

36. Haas, *Tribal Government Under I.R.A.*, pp. 29–30.

37. Foley, "Origins of the IRA," p. 72.

38. Haas, *Tribal Government Under I.R.A.*, p. 14; Philp, *Collier's Crusade*, p. 163.

39. Theodore H. Haas, "The Indian Reorganization Act in Historical Perspective," in Kelly, ed., *Indian Affairs*, p. 19; U.S., *Statutes at Large* 64 (April 19, 1950): 44 [Navajo–Hopi Rehabilitation Act].

40. Haas, *Tribal Government Under I.R.A.*, pp. 11, 32, 34.

41. Bromert, "Sioux and the Indian New Deal," App. A, p. 199.

42. Ducheneaux, "IRA and the Cheyenne River Sioux," p. 12.

43. Ibid., p. 10.

44. Ibid., p. 13.

45. Ibid.

46. Clarence Wesley, "Tribal Self-Government Under the IRA," in Kelly, ed., *Indian Affairs*, p. 26.

47. Bromert, "Sioux and the Indian New Deal," pp. 76–77.

48. Walter F. Dickens to John Collier, May 10, 1940, Cheyenne River Agency IRA, Richardson Archives, University of South Dakota, Vermillion, quoted in Bromert, "Sioux and the Indian New Deal," p. 84.

49. Ibid., pp. 75–79.

50. McNickle, "Indian New Deal," p. 115.

51. "Is This Real," in *The Portable North American Indian Reader*, ed. Frederick W. Turner III (New York; Viking, 1973), p. 241.

Chapter 5

1. Marcella Cash, "The Indian Reorganization Act on the Rosebud Reservation" (M.A. thesis, University of South Dakota, 1976), pp. 67–68, as discussed in Roger Bromert, "The Sioux and the Indian New Deal, 1933–1945" (Ph.D. diss., University of Toledo, 1980), pp. 177–178.

2. Kenneth R. Philp, *John Collier's Crusade for Indian Reform, 1920–1954* (Tucson: University of Arizona Press, 1977), pp. 194–197.

3. *United States v. Klamath Indians* 304 U.S. 119 (1938). See also Robert W. Oliver, "The Legal Status of American Indian Tribes," *Oregon Law Review* 38 (April 1959): 204–205.

4. "The American Indian—Tribal Sovereignty and Civil Rights," *Iowa Law Review* 51 (Spring 1966): 654–669.

5. Jerry L. Bean, "The Limits of Indian Tribal Sovereignty: The Cornucopia of Inherent Powers," *North Dakota Law Review* 49 (Winter 1973): 306; Heinrich Krieger, "Principles of the Indian Law and the Act of June 18, 1934," *George Washington University Law Review* 3 (March 1935): 290; Ben Cashman, "The American Indian—Standing in a Peculiar Legal Relation" (Ph.D. diss., University of Washington, 1969), p. 114.

6. Philp, *Collier's Crusade*, p. 120. The most recent edition of Cohen's pathbreaking work is Rennard Strickland and Charles F. Wilkinson, eds., *Felix S. Cohen's Handbook of Federal Indian Law* (Charlottesville, Va.: Michie Bobbs-Merrill, 1982).

7. Bean, "Limits of Indian Tribal Sovereignty," p. 312; Oliver, "Legal Status of American Indian Tribes," p. 229.

8. Philp, *Collier's Crusade,* p. 170.

9. Henry F. Dobyns, "The Indian Reorganization Act and Federal Withdrawal," *Applied Anthropology* 7 (Spring 1948): 40.

10. Ibid., p. 41.

11. Theodore W. Taylor, *American Indian Policy* (Mount Airy, Md.: Lomond, 1983), p. 10.

12. Ibid. See also Chapter 9; *Santa Clara Pueblo v. Martinez* 436 U.S. 49 (1978).

13. *Superintendent of the Five Civilized Tribes v. Commissioner* 295 U.S. 418 (1935).

14. U.S., *Statutes at Large* 47 (June 6, 1932): 169–289; sec. 624, Exemption of Articles manufactured or produced by Indians, p. 269.

15. *United States v. Charles* 23 F. Supp. 346 (1938); *United States v. Forness* 125 F. 928 (1942). For a discussion of this particular issue, see Oliver, "Legal Status of American Indian Tribes," p. 214.

16. See Constitution, Omaha Tribe of Nebraska, art. IV, sec. 1(h), and Constitution, Fort McDermitt Paiute and Shoshone Tribe, art. VI, sec. 1(f), as noted in Strickland and Wilkinson, eds., *Cohen's Indian Law,* p. 267.

17. Dobyns, "Indian Reorganization Act," p. 39.

18. Strickland and Wilkinson, eds., *Cohen's Indian Law,* p. 157.

19. Ibid., pp. 157–159.

20. *Porter v. Hall* 34 Ariz. 308 (1928); N. D. Houghton, *"Wards of the United States"—Arizona Applications: A Study of the Legal Status of Indians* (Tucson: University of Arizona Press, 1946), pp. 5, 19; William Hughes, "Indians on a New Trail," *Catholic World* 139 (July 1934): 467.

21. Rudolph X. Foley, "The Origins of the Indian Reorganization Act of 1934" (Ph.D. diss., Fordham University, 1937), pp. 198–199; Lawrence C.

Kelly, "John Collier and the Indian New Deal: An Assessment," in *Indian–White Relations: A Persistent Paradox,* ed. Jane F. Smith and Robert M. Kvasnicka (Washington, D.C.: Howard University Press, 1976), p. 238.

22. Clarence Wesley, "Tribal Self-Government Under the IRA," in *Indian Affairs and the Indian Reorganization Act: The Twenty Year Record,* ed. William H. Kelly (Tucson: University of Arizona, 1954), p. 27; Donald L. Parman, "The Indian and the Civilian Conservation Corps," *Pacific Historical Review* 40 (February 1971): 43–45; Roger Bromert, "The Sioux and the Indian-CCC," *South Dakota History* 8 (Fall 1978): 350; Lawrence C. Kelly, "The Indian Reorganization Act: The Dream and the Reality," *Pacific Historical Review* 41 (August 1975): 293–312.

23. Dobyns, "Indian Reorganization Act," p. 38; D'Arcy McNickle, "The American Indian Today," *Missouri Archaeologist* 5 (September 1939): 8; Philp, *Collier's Crusade,* pp. 187–190; D'Arcy McNickle, "The Indian New Deal as Mirror of the Future," in *Political Organization of Native North Americans,* ed. Ernest L. Schusky (Washington, D.C.: University Press of America, 1980), pp. 112–113.

24. Bromert, "Sioux and the Indian New Deal," p. 58; Kelly, "Indian Reorganization Act," pp. 311–312; Philp, *Collier's Crusade,* p. 205.

25. U.S., *Statutes at Large* 48 (June 18, 1934): 984–988 [Indian Reorganization Act or Wheeler–Howard Act].

26. Curtis Berkey, "The Legislative History of the Indian Reorganization Act," *American Indian Journal* 2 (July 1976): 21.

27. U.S., *Statutes at Large* 60 (August 13, 1946): 1049–1056 [Indian Claims Commission Act].

28. Jay B. Nash, ed., *The New Day for the Indians: A Survey of the Working of the Indian Reorganization Act of 1934* (New York: Academy Press, 1938), p. 39.

29. Cashman, "American Indian," p. 164.

30. U.S., *Statutes at Large* 12 (March 3, 1863): 765–768.

31. Thomas Le Duc, "The Work of the Indian Claims Commission Under the Act of 1946," *Pacific Historical Review* 26 (February 1957): 1–2.

32. Nancy Oesterich Lurie, "The Indian Claims Commission Act," *Annals of the American Academy of Political and Social Science* 311 (May 1957): 56–57.

33. Ibid., p. 57. Similar, but not exactly the same, figures are given in Francis Paul Prucha, *The Great Father: The United States Government and the American Indians,* 2 vols. (Lincoln: University of Nebraska Press, 1984), vol. 2, p. 1018.

34. *United States v. Creek Nation* 295 U.S. 103 (1935).

35. *Shoshone Tribe of Indians v. United States* 299 U.S. 476 (1937).

36. *Seminole Nation v. United States* 316 U.S. 286 (1941); *Seminole Nation v. United States* 316 U.S. 310 (1942).

37. U.S., *Public Papers of the Presidents of the United States: Harry S. Truman, 1946* (Washington, D.C.: Government Printing Office, 1962), p. 414, quoted in Prucha, *Great Father,* vol. 2, p. 1019.

38. U.S., Indian Claims Commission Act, pp. 1050–1055.

39. Ibid., pp. 1050, 1052.

40. Ibid., p. 1050.

41. Ibid.

42. Lurie, "Indian Claims Commission Act," pp. 58–61.

43. For a general treatment, see Graham D. Taylor, *The New Deal and American Indian Tribalism: The Administration of the Indian Reorganization Act, 1934–1945* (Lincoln: University of Nebraska Press, 1980).

44. Bromert, "Sioux and the Indian New Deal," pp. 89–96.

45. Foley, "Origins of the IRA," p. 166; Houghton, *"Wards of the United States,"* p. 15; McNickle, "Indian Today," p. 3; Theodore H. Haas, "The Indian Reorganization Act in Historical Perspective," in Kelly, ed., *Indian Affairs,* p. 18; John Collier, "Collier Replies to Mekeel," *American Anthropologist* 46 (July–September 1944): 423; Philp, *Collier's Crusade,* pp. 176, 185.

46. Krieger, "Principles of the Indian Law," p. 304.

47. U.S., Indian Reorganization Act, pp. 984–988.

48. Ibid.; Nash, ed., *New Day for the Indians,* p. 33.

49. Kenneth R. Philp, "John Collier and the Controversy over the Wheeler–Howard Bill," in Smith and Kvasnicka, eds., *Indian–White Relations,* pp. 178, 189; Foley, "Origins of the IRA," pp. 157–159; Collier, "Replies to Mekeel," p. 426.

50. McNickle, "Indian New Deal," p. 111.

51. Wesley, "Tribal Self-Government," p. 28.

52. Quoted in Bromert, "Sioux and the Indian-CCC," p. 342.

53. McNickle, "American Indian Today," p. 2.

Chapter 6

1. Charles F. Wilkinson and Eric R. Biggs, "Evolution of the Termination Policy," *American Indian Law Review* 5, no. 1 (1977): 140, 166, n.3.

2. Ibid., pp. 140–145.

3. Vine Deloria, Jr., and Clifford M. Lytle, *American Indians, American Justice* (Austin: University of Texas Press, 1983), pp. 16–17; Larry W. Burt, *Tribalism in Crisis: Federal Indian Policy, 1953–1961* (Albuquerque: University of New Mexico Press, 1982), pp. 4–6.

4. U.S., *Statutes at Large* 67 (August 1, 1953): B132 ["Indians," House Concurrent Resolution 108].

5. Ibid.

6. Russel Lawrence Barsh and James Youngblood Henderson, *The Road: Indian Tribes and Political Liberty* (Berkeley: University of California Press, 1980), p. 126.

7. Arthur V. Watkins, "Termination of Federal Supervision: The Removal of Restrictions over Indian Property and Person," *Annals of the American Academy of Political and Social Science* 311 (May 1957): 55.

8. Donald L. Fixico, *Termination and Relocation: Federal Indian Policy, 1945–1960* (Albuquerque: University of New Mexico Press, 1986), p. 103; Burt, *Tribalism in Crisis*, pp. 29–30.

9. Burt, *Tribalism in Crisis*, pp. 30–46.

10. Wilkinson and Biggs, "Evolution of Termination," p. 151.

11. Stephen Herzberg, "The Menominee Indians: Termination to Restoration," *American Indian Law Review* 6, no. 1 (1978): 158–160.

12. Burt, *Tribalism in Crisis*, p. 42; Nancy Oestreich Lurie, "Menominee Termination: From Reservation to Colony," *Human Organization* 31 (Fall 1972): 261.

13. Lurie, "Menominee Termination," pp. 257–270.

14. Burt, *Tribalism in Crisis*, p. 91.

15. Ibid., pp. 107–108.

16. Wilkinson and Biggs, "Evolution of Termination," pp. 153–154.

17. Fixico, *Termination and Relocation*, pp. 183–185.

18. Ibid., pp. 134–136.

19. Ruth Mulvey Harmer, "Uprooting the Indians," *Atlantic Monthly,* March 1956, pp. 54–57. See also Fixico, *Termination and Relocation*, pp. 148–150, 155, 169.

20. Burt, *Tribalism in Crisis*, pp. 56–58.

21. Ibid., p. 73.

22. Dorothy Van de Mark, "The Raid on the Reservations," *Harper's Magazine,* March 1956, pp. 48–53.

23. U.S., *Statutes at Large* 67 (August 15, 1953): 588–590 [Public Law 280].

24. Vine Deloria, Jr., "Legislation and Litigation Concerning American Indians," *Annals of the American Academy of Political and Social Science* 436 (March 1978): 96.

25. Public Law 280, pp. 589–590.

26. Ibid., sec. 2b, p. 589.

27. Ibid., sec. 4c.

28. Barsh and Henderson, *The Road*, p. 127.

29. Frederick J. Stefon, "The Irony of Termination: 1943–1958," *Indian Historian* 11 (Summer 1978): 7.

30. Wilkinson and Biggs, "Evolution of Termination," p. 158.

31. Stefon, "Irony of Termination," p. 9.

32. Van de Mark, "Raid on the Reservations," p. 51; Harmer, "Uprooting the Indians," p. 54.

33. Burt, *Tribalism in Crisis,* p. 77.

34. Quoted in Carole E. Goldberg, "Public Law 280: The Limits of State Jurisdiction over Reservation Indians," *UCLA Law Review* 22 (February 1975): 552–553, esp. n. 92.

35. Ibid., pp. 546–547, esp. nn.56, 57.

36. Ibid., p. 548, n.72; Montana, *Montana Revised Code Annotated,* sec. 83–806 (1966).

37. *Williams v. Lee* 358 U.S. 217 (1959).

38. Ibid., p. 223.

39. *Kake v. Egan* 369 U.S. 60 (1962).

40. Vine Deloria, Jr., and Clifford M. Lytle, *The Nations Within: The Past and Future of American Indian Sovereignty* (New York: Pantheon Books, 1984), p. 191.

41. Stefon, "Irony of Termination," p. 5; Raymond V. Butler, "The Bureau of Indian Affairs: Activities Since 1945," *Annals of the American Academy of Political and Social Science* 436 (March 1978): 52; John R. White, "Barmecide Revisited: The Gratuitous Offset in Indian Claims Cases," *Ethnohistory* 25 (Spring 1978): 180.

42. Fixico, *Termination and Relocation,* p. 30.

43. Ibid.; Thomas LeDuc, "The Work of the Indian Claims Commission Under the Act of 1946," *Pacific Historical Review* 26 (February 1957): 2.

44. Nancy Oestreich Lurie, "Indian Claims Commission," *Annals of the American Academy of Political and Social Science* 436 (March 1978): 99; Nancy Oestreich Lurie, "The Indian Claims Commission Act," *Annals of the American Academy of Political and Social Science* 311 (May 1957): 68–69.

45. Watkins, "Termination of Federal Supervision," p. 50.

46. Lurie, "Indian Claims Commission," p. 107.

47. LeDuc, "Work of the ICC," pp. 7–15; Ralph A. Barney, "Legal Problems Peculiar to Indian Claims Litigation," *Ethnohistory* 2 (Fall 1955): 320–325; Lurie, "Indian Claims Commission," pp. 97–110; Lurie, "Indian Claims Act," pp. 56–70.

48. *Choctaw Nation v. United States* 1 Ind. Cl. Comm. 182 (1950).

49. LeDuc, "Work of the ICC," pp. 5–6, 13.

50. Herbert T. Hoover, "Yankton Sioux Tribal Claims Against the United States, 1917–1975," *Western Historical Quarterly* 7 (April 1976): 137–138; White, "Barmecide Revisited," pp. 179–184.

51. LeDuc, "Work of the ICC," p. 15.

52. Berlin B. Chapman, "The Day in Court for the Kiowa, Comanche and Apache Tribes," *Great Plains Journal* 2 (Fall 1962): 1–21.

53. Deloria and Lytle, *Nations Within*, p. 261.

54. *Tee-Hit-Ton Indians v. United States* 348 U.S. 273 (1955).

55. Ibid., pp. 273–277.

56. Ibid., p. 277.

57. *Beecher v. Wetherby* 95 U.S. 517 at 525 (1877).

58. *Tee-Hit-Ton Indians v. United States* 348 U.S. 273 at 281 (1955). See also *United States v. Alcea Band of Tillamooks* 341 U.S. 48 (1951).

59. Barsh and Henderson, *The Road*, pp. 140–143. See also Harvey D. Rosenthal, "Indian Claims and the American Conscience: A Brief History of the Indian Claims Commission," in *Irredeemable America: The Indians' Estate and Land Claims*, ed. Imre Sutton (Albuquerque: University of New Mexico Press, 1985), pp. 52–53; David H. Getches, "Alternative Approaches to Land Claims: Alaska and Hawaii," in Sutton, ed., *Irredeemable America*, pp. 303–305.

60. *Otoe and Missouria Tribe of Indians v. United States* 131 Ct. Cl. 593 (1955). See also Monroe E. Price, *Law and the American Indian: Readings, Notes, and Cases* (Indianapolis: Bobbs-Merrill, 1973), pp. 470–476.

61. John R. Wunder, *The Kiowa* (New York: Chelsea House, 1989), pp. 80–93.

62. Deloria and Lytle, *American Indians*, pp. 230–234.

63. *Native American Church v. Navajo Tribal Council* 272 F.2d 131 (1959).

64. *People v. Woody* 61 Cal. 2d 716 (1964).

65. *Squire v. Capoeman* 351 U.S. 1 (1956).

66. *Holt v. Commissioner* 364 F.2d 38 (1966).

67. *Warren Trading Post Co. v. Arizona Tax Commission* 380 U.S. 685 (1965).

68. The Supreme Court ruled during its January 1992 session that Indian reservation land privately owned by Native Americans can be taxed by states and counties ("U.S. High Court: Privately Held Indian Land Can Be Taxed," *Lincoln [Nebraska] Star*, January 15, 1992).

69. See pp. 51–54.

70. *Arizona v. California* 373 U.S. 546 (1963). See also Lloyd Burton, *American Indian Water Rights and the Limits of Law: Reflections in a Glass Bead* (Lawrence: University Press of Kansas, 1991).

71. *Colliflower v. Garland* 342 F.2d 369 (1965). See also a future, although limited, application in *Settler v. Yakima Tribal Court* 419 F.2d 486 (1969).

72. Fixico, *Termination and Relocation*, pp. 190–199.

73. Burt, *Tribalism in Crisis*, pp. 118–123.

74. Linda Hogan, *Mean Spirit: A Novel* (New York: Atheneum, 1990), p. 219.

75. Hank Adams, in *Indian Self-Rule: First-Hand Accounts of Indian–White*

Relations from Roosevelt to Reagan, ed. Kenneth R. Philp (Salt Lake City: Howe Brothers, 1986), p. 239.

Chapter 7

1. Crazy Horse, ''We Preferred Our Own Way of Living,'' in *Indian Oratory: Famous Speeches by Noted Indian Chieftains,* comp. W. C. Vanderwerth (New York: Ballantine Books, 1972), p. 175.

2. Donald L. Burnett, Jr., ''An Historical Analysis of the 1968 'Indian Civil Rights' Act,'' *Harvard Journal of Legislation* 9 (May 1972): 574–576.

3. Ibid., p. 575, quoting a letter from Lawrence M. Baskir, chief counsel and staff director, Subcommittee on Constitutional Rights of the Senate Committee on the Judiciary, to Donald L. Burnett Jr., March 5, 1970. See also Paul R. Clancy, *Just a Country Lawyer: A Biography of Senator Sam Ervin* (Bloomington: Indiana University Press, 1974), pp. 203–207.

4. Arthur Lazarus, Jr., ''Title II of the 1968 Civil Rights Act: An Indian Bill of Rights,'' *North Dakota Law Review* 45 (Spring 1969): 338.

5. Lyndon B. Johnson, ''The Forgotten American'' (Presidential message delivered to Congress, March 6, 1968), quoted in James E. Officer, ''The Bureau of Indian Affairs Since 1945: An Assessment,'' *Annals of the American Academy of Political and Social Science* 436 (March 1978): 67.

6. Robert L. Bennett, ''Problem and Prospects in Developing Indian Communities,'' *Arizona Law Review* 10 (Winter 1968): 655.

7. Ibid., pp. 657–659; Raymond V. Butler, ''The Bureau of Indian Affairs: Activities Since 1945,'' *Annals of the American Academy of Political and Social Science* 436 (March 1978): 57.

8. *Menominee Tribe of Indians v. United States* 391 U.S. 404 (1968).

9. *United States v. Northern Paiute Nation et al.* 393 F.2d 786 (1968); *Gila River Pima–Maricopa Indian Community et al. v. United States* (Gila River I) 20 Ind. Cl. Comm. 131 (1968); *Peoria Tribe of Indians of Oklahoma et al. v. United States* 390 U.S. 468 (1968).

10. *Fournier v. Roed* 161 N.W.2d 458 (1968); *Boyer v. Shoshone–Bannock Indian Tribes* 441 P.2d 167 (1968); *Pourier v. Board of County Commissioners* 157 N.W.2d 532 (1968); *Makah Indian Tribe v. Clallam County* 440 P.2d 442 (1968).

11. *Puyallup Tribe, Inc. v. Department of Game* (Puyallup I) 391 U.S. 392 (1968); *People v. Jondreau* 166 N.W.2d 293 (1968).

12. *Dodge v. Nakai* (Dodge I) 298 F. Supp. 17 (1968).

13. Robert C. Carriker, ''The Kalispel Tribe and the Indian Claims Commission Experience,'' *Western Historical Quarterly* 9 (January 1978): 31, n.38.

14. *Peoria Tribe of Indians of Oklahoma et al. v. United States* 390 U.S. 468 (1968).

15. *Gila River Pima–Maricopa Indian Community et al. v. United States* (Gila River II) 427 F.2d 1194 (1970).

16. *United States v. Northern Paiute Nation et al.* 393 F.2d 786 (1968).

17. Frederick J. Stefon, "The Irony of Termination: 1943–1958," *Indian Historian* 11 (Summer 1978): 11.

18. Earl Old Person, "Statement of Earl Old Person, Chairman of the Blackfeet Tribe of Montana" (Presented at the conference of the Bureau of Indian Affairs, Spokane, Washington, 1966), in *Of Utmost Good Faith,* comp. Vine Deloria, Jr. (San Francisco: Straight Arrow Books, 1971), p. 219.

19. *Menominee Tribe of Indians v. United States* 391 U.S. 404 (1968).

20. Ibid., pp. 410–411. See also, for a brief analysis, "Implication of Civil Remedies Under the Indian Civil Rights Act," *Michigan Law Review* 75 (November 1976): 219–221.

21. Charles F. Wilkinson and Eric R. Biggs, "Evolution of the Termination Policy," *American Indian Law Review* 5, no. 1 (1977): 153–154, n.144.

22. Carole E. Goldberg, "Public Law 280: The Limits of State Jurisdiction over Reservation Indians," *UCLA Law Review* 22 (February 1975): 559–560.

23. Edward J. Ward, "Minority Rights and American Indians," *North Dakota Law Review* 51 (Fall 1974): 138–139.

24. U.S., *Statutes at Large* 82 (April 11, 1968): 80 [Addition to Major Crimes Act]; G. Kenneth Reiblich, "Indian Rights Under the Civil Rights Act of 1968," *Arizona Law Review* 10 (Winter 1968): 642.

25. Lazarus, "Title II," p. 341; Albert E. Kane, "The Negro and the Indian: A Comparison of Their Constitutional Rights," *Arizona Law Review* 7 (Spring 1966): 246–247; Albert E. Kane, "Indians—Criminal Procedure: Habeas Corpus as an Enforcement Procedure Under the Indian Civil Rights Act of 1968," *Washington Law Review* 46 (May 1971): 544–545; Burnett, "Historical Analysis," p. 579.

26. Lazarus, "Title II," p. 345.

27. Burnett, "Historical Analysis," p. 587; James R. Kerr, "Constitutional Rights, Tribal Justice, and the American Indian," *Journal of Public Laws* 18 (1969): 326.

28. Burnett, "Historical Analysis," pp. 584–588; Reiblich, "Indian Rights," p. 634.

29. *Talton v. Mayes* 163 U.S. 376 (1896).

30. *Colliflower v. Garland* 342 F.2d 369 (1965). Other cases generating considerable discussion but not going as far as *Colliflower* in overruling the *Talton* Doctrine are *Barta v. Oglala Sioux Tribe* 259 F.2d 553 (1958); *Glover v. United States* 219 F. Supp. 19 (1963); *Twin Cities Chippewa Tribal Council v.*

Minnesota Chippewa Tribe 370 F.2d 529 (1967); *Dicke v. Cheyenne–Arapaho Tribes, Inc.* 304 F.2d 113 (1962); *Prairie Band of Pottawatomie Tribe v. Puckkee* 321 F.2d 767 (1963); and *Iron Crow v. Oglala Sioux Tribe* 231 F.2d 89 (1956).

31. Joseph de Raismes, "The Indian Civil Rights Act of 1968 and the Pursuit of Responsible Tribal Self-Government," *South Dakota Law Review* 20 (Winter 1975): 65. For further discussions on the legalisms of *Colliflower v. Garland,* see John R. White, "Civil Rights and the Native American," *Integrated Education* 11 (November–December 1973): 32–33; John R. White, "The Indian Bill of Rights and the Constitutional Status of Tribal Governments," *Harvard Law Review* 82 (April 1969): 1348–1350; Lazarus, "Title II," pp. 343–344; and Burton D. Fretz, "The Bill of Rights and American Indian Tribal Governments," *Natural Resources Journal* 6 (October 1966): 585.

32. *Toledo v. Pueblo de Jemez* 119 F. Supp. 429 (1954). For discussions of this case, see Kane, "Negro and the Indian," p. 247; "Indian Tribes and Civil Rights," *Stanford Law Review* 7 (March 1955): 285–287; and Kerr, "Constitutional Rights," p. 325. Vine Deloria, Jr., believes that the religious freedom cases were mainly responsible for the passage of the Indian Bill of Rights ("Implications of the 1968 Civil Rights Act in Tribal Autonomy," in Deloria, *Indian Voices: The First Convocation of American Indian Scholars* [San Francisco: Indian Historian Press, 1970], pp. 85–87).

33. *Native American Church v. Navajo Tribal Council* 272 F.2d 131 (1959). See also Kerr, "Constitutional Rights," pp. 324–325.

34. Donald L. Fixico, *Termination and Relocation: Federal Indian Policy, 1945–1960* (Albuquerque: University of New Mexico Press, 1986), p. 199.

35. U.S. Congress, Senate, S. 3047, 88th Cong., 2nd sess., 1964, quoted in "The Constitutional Rights of the American Tribal Indian," *Virginia Law Review* 51 (January 1965): 137, also pp. 137–140.

36. Michael Smith, "Tribal Sovereignty and the 1968 Indian Bill of Rights," *Civil Rights Digest* 3 (Summer 1970): 12–13.

37. U.S., *Statutes at Large* 82 (April 11, 1968): 73–92.

38. Ibid., p. 77.

39. Fixico, *Termination and Relocation,* p. 200.

40. "Robert Burnette," in *Indian Self-Rule: First-Hand Accounts of Indian–White Relations from Roosevelt to Reagan,* ed. Kenneth R. Philp (Salt Lake City: Howe Brothers, 1986), pp. 105–106.

41. Vine Deloria, Jr., "Legislation and Litigation Concerning American Indians," *Annals of the American Academy of Political and Social Science* 436 (March 1978): 91.

42. Helen M. Scheirbeck, "Federal Indian Policy, 1960–1976," in Philp, ed., *Indian Self-Rule,* pp. 216–217.

43. Quoted in Smith, "Tribal Sovereignty," p. 14.

44. Ibid., pp. 14–15. See also Ernest L. Schusky, "American Indians and the 1968 Civil Rights Act," *America Indigena* 29 (April 1969): 372–374.

45. Gerald Wilkinson, quoting the chairman of the Mescalero Apache Tribe (Minutes of the meeting of the A.C.L.U. Indian Rights Commission, August 9–10, 1974), quoted in de Raismes, "Indian Civil Rights Act," p. 59.

46. Ibid., pp. 59–106; Michael R. Granen and Dougles E. Somers, "Indian Bill of Rights," *Southwestern University Law Review* 5 (Spring 1973): 163–164; John S. Warren, "An Analysis of the Indian Bill of Rights," *Montana Law Review* 33 (Summer 1972): 263–265; Kerr, "Constitutional Rights," pp. 332–338.

47. Lazarus, "Title II," pp. 346–347.

48. The facts of the case are found in two opinions: *Dodge v. Nakai* 298 F. Supp. 17 (1968) and *Dodge v. Nakai* (Dodge II) 298 F. Supp. 26 (1969).

49. *Dodge v. Nakai* 298 F. Supp. 17 at 24 (1968).

50. *Dodge v. Nakai* 298 F. Supp. 26 at 31–32 (1969).

51. For discussions on the *Dodge* cases, see Reiblich, "Indian Rights," pp. 624–625; Dennis R. Holmes, "Political Rights Under the Indian Civil Rights Act," *South Dakota Law Review* 24 (Spring 1979): 419–446; Alvin J. Ziontz, "In Defense of Tribal Sovereignty: An Analysis of Judicial Error in Construction of the Indian Civil Rights Act," *South Dakota Law Review* 20 (Winter 1975): 1–58; Kerr, "Constitutional Rights," pp. 320–321; "Implication of Civil Remedies Under the Indian Civil Rights Act," pp. 210–235; Granen and Somers, "Indian Bill of Rights," pp. 139–143; and Cliff A. Jones, "Remedies: Tribal Deprivation of Civil Rights: Should Indians Have a Cause of Action Under 42 U.S.C. 1983?" *American Indian Law Review* 3, no. 1 (1975): 183–195.

52. Janet Campbell Hale, *The Jailing of Cecelia Capture* (Albuquerque: University of New Mexico Press, 1985), p. 67.

Chapter 8

1. J. Dineley Prince, "A Passamaquoddy Tobacco Famine," *International Journal of American Linguistics* 1 (July 1917): 58–63.

2. Quoted in Michael Smith, "Tribal Sovereignty and the 1968 Indian Bill of Rights," *Civil Rights Digest* 3 (Summer 1970): 14.

3. Alvin J. Ziontz, "In Defense of Tribal Sovereignty: An Analysis of Judicial Error in Construction of the Indian Civil Rights Act," *South Dakota Law Review* 20 (Winter 1975): 2–3.

4. Mary L. Muehlen, "An Interpretation of the Due Process Clause of the Indian Bill of Rights," *North Dakota Law Review* 51 (Fall 1974): 203–204.

5. Judy D. Lynch, "Indian Sovereignty and Judicial Interpretations of the

Indian Civil Rights Act,'' *Washington University Law Quarterly* (Summer 1979): 912.

6. U.S., Congress, Senate, *Hearings on Constitutional Rights of the American Indian Before the Subcommittee on Constitutional Rights of the Senate Committee on the Judiciary,* 89th Cong., 1st sess., 1966, p. 196, quoted in Jennifer B. Beaver, ''Political Advocacy and Freedom of Expression Under the Indian Civil Rights Act of 1968,'' *Arizona State Law Journal* (1976): 487 n.47.

7. *Dodge v. Nakai* 298 F. Supp. 17 (1968).

8. *Janis v. Wilson* 521 F.2d 724 (1975).

9. *White Eagle v. One Feather* 478 F.2d 1314 (1973), quoted in Gary D. Kennedy, ''Tribal Elections: An Appraisal After the Indian Civil Rights Act,'' *American Indian Law Review* 3, no. 2 (1975): 498.

10. *Daly v. United States* 483 F.2d 700 (1973). See also *Brown v. United States* 486 F.2d 658 (1973).

11. *Luxon v. Rosebud Sioux Tribe* 455 F.2d 698 (1972).

12. *McCurdy v. Steele* 506 F.2d 653 (1973).

13. *Daly v. United States* 483 F.2d 700 (1973); *Jacobson v. Forest County Potawatomi Community* 389 F. Supp. 994 (1974). See also *Howlett v. Salish & Kootenai Tribes* 529 F.2d 233 (1976); Ralph W. Johnson and E. Susan Crystal, ''Indians and Equal Protection,'' *Washington Law Review* 54 (June 1979): 623–625; *Groundhog v. Keeler* 442 F.2d 674 (1971); Joseph de Raismes, ''Indian Civil Rights Act of 1968 and the Pursuit of Responsible Tribal Self-Government,'' *South Dakota Law Review* 20 (Winter 1975): 59–106.

14. *Jacobson v. Forest County Potawatomi Community* 389 F. Supp. 994 (1974); *Wounded Head v. Tribal Council of the Oglala Sioux Tribe* 507 F.2d 1079 (1975).

15. *Williams v. Sisseton–Wahpeton Sioux Tribal Council* 387 F. Supp. 1194 (1975).

16. *Means v. Wilson* 383 F. Supp. 378 (1974).

17. *Solomon v. LaRose* 335 F. Supp. 715 (1971).

18. *Morton v. Mancari* 417 U.S. 535 (1974); Johnson and Crystal, ''Indians and Equal Protection,'' pp. 595–596.

19. *O'Neal v. Cheyenne River Sioux Tribe* 482 F.2d 1140 (1973).

20. *United States ex rel. Cobell v. Cobell* 503 F.2d 700 (1974).

21. *Santa Clara Pueblo v. Martinez* 436 U.S. 49 (1978), nn.2, 3; Fred Martinez, ''Indian Women and Tribal Enrollment: The Case of *Martinez v. Santa Clara Pueblo*'' (Paper presented at the annual meeting of the Western History Association, Austin, Texas, October 18, 1991).

22. *Santa Clara Pueblo v. Martinez* 436 U.S. 49 at 53–54 (1978).

23. Ibid., pp. 54–55.

24. Ibid., p. 55.

25. Ibid.

26. Ibid., p. 71.

27. Ibid., p. 58.

28. Ibid., p. 64.

29. For a discussion of the implications of *Martinez,* see Johnson and Crystal, "Indians and Equal Protection," pp. 627–636; Andra Pearldaughter, "Constitutional Law: Equal Protection: *Martinez v. Santa Clara Pueblo*—Sexual Equality Under the Indian Civil Rights Act," *American Indian Law Review* 6, no. 1 (1978): 187–204; John R. Hardin, *"Santa Clara Pueblo v. Martinez:* Tribal Sovereignty and the Indian Civil Rights Act of 1968," *Arkansas Law Review* 33 (Summer 1979): 399–421; Vieno Lindstrom, "Constitutional Law: *Santa Clara Pueblo v. Martinez:* Tribal Membership and the Indian Civil Rights Act," *American Indian Law Review* 6, no. 1 (1978): 205–216; Susan Sanders Molander, "Indian Civil Rights Act and Sex Discrimination—*Martinez v. Santa Clara Pueblo,"* *Arizona State Law Journal* (1977): 227–239; and Alvin J. Ziontz, "After *Martinez:* Civil Rights Under Tribal Government," *University of California–Davis Law Review* 12 (March 1979): 1–35.

The controversy, however, continues for the Santa Clara Pueblo. In 1991, a Santa Clara Pueblo Tribal Council decision authorized the governor to set up a committee to revise the tribal census, and a new uproar is being heard among the people (Martinez, "Indian Women and Tribal Enrollment").

30. James S. Olson and Raymond Wilson, *Native Americans in the Twentieth Century* (Urbana: University of Illinois Press, 1984), pp. 159–169.

31. Peter Blue Cloud, ed., *Alcatraz Is Not An Island* by Indians Of All Tribes (Berkeley, Calif.: Wingbow Press, 1972), p. 43.

32. Lyndon Johnson, "The Forgotten American" (Presidential message delivered to Congress, March 6, 1968), discussed in James E. Officer, "The Bureau of Indian Affairs Since 1945: An Assessment," *Annals of the American Association of Political and Social Science* 436 (March 1978): 66; Rebecca L. Robbins, "The Forgotten American: A Foundation for Contemporary American Indian Self-Determination," *Wicazo Sa Review* 6 (Spring 1990): 31–32.

33. *Indian Record* (August 1970), p. 3, quoted in Raymond V. Butler, "The Bureau of Indian Affairs: Activities Since 1945," *Annals of the American Association of Political and Social Science* 436 (March 1978): 57.

34. Emma R. Gross, *Contemporary Federal Policy Toward Indians* (Westport, Conn.: Greenwood Press, 1989), pp. 70–71.

35. Richard Nixon, "Indian Message to the Congress," *Congressional Quarterly Almanac* (1970), p. 102–A, quoted in Gross, *Contemporary Federal Policy,* p. 35.

36. U.S., *Statutes at Large* 88 (January 4, 1975): 2203–2217 [Indian Self-Determination and Education Act].

37. U.S., *Statutes at Large* 88 (January 2, 1975): 1910–1914 [Indian Policy Review Commission Resolution].

38. U.S., Indian Self-Determination and Education Act, p. 2203.

39. U.S., Indian Policy Review Commission Resolution, pp. 1910–1911.

40. Ibid., p. 1910; Gross, *Contemporary Federal Policy,* p. 41.

41. U.S., Indian Policy Review Commission Resolution, pp. 1911–1912.

42. Officer, "Bureau of Indian Affairs Since 1945," p. 71; U.S., American Indian Policy Review Commission, *Final Report* (Washington, D.C.: Government Printing Office, 1977).

43. American Indian Policy Review Commission, *Final Report,* p. 574, quoted in Gross, *Contemporary Federal Policy,* p. 43.

44. Richard Real Bird, chairman, Crow Nation, "Crow Sovereignty" (Comments presented at "The Crow Nation: A Historical and Cultural Seminar," Billings, Montana, June 23, 1989).

45. Olson and Wilson, *Native Americans,* pp. 17, 138, 195–196.

46. Quoted in Gross, *Contemporary Federal Policy,* p. 67.

47. Ibid., pp. 66–68.

48. The term "Eskimo" is found in many texts and in congressional statutes and debates. This word is not preferred by many of Alaska's and Canada's native peoples, who frequently use "Inuit" to describe themselves collectively.

49. Sidney L. Harring, "The Incorporation of Alaskan Natives Under American Law: United States and Tlingit Sovereignty, 1867–1900," *Arizona Law Review* 31, no. 2 (1989): 279–280.

50. U.S., *Statutes at Large* 92 (December 18, 1971): 688–715 [Alaska Native Claims Settlement Act].

51. Gross, *Contemporary Federal Policy,* p. 26.

52. Arthur Lazarus, Jr., and W. Richard West, Jr., "The Alaska Native Claims Settlement Act: A Flawed Victory," *Law and Contemporary Problems* 11 (Winter 1976): 132–165; Karen Perret, "The Alaska Native Claims Settlement Act and the Alaskans," *Indian Historian* 11, no. 1 (1978): 3–9; Kerry Stoebner, Vicki Camerino, and Steve Nickeson, "Alaska Native Water Rights as Affected by the Alaska Native Claims Settlement Act," *American Indian Journal* 4 (March 1978): 2–26; Monroe E. Price, Richard R. Purtich, and D. Gerber, "The Tax Exemption of Native Lands Under Section 21(d) of the Alaska Native Claims Settlement Act," *UCLA–Alaska Law Review* 6 (Fall 1976): 1–33; Monroe E. Price, "Region–Village Relations Under the Alaska Native Claims Settlement Act," *UCLA–Alaska Law Review* 5 (Fall 1975): 58–79.

53. See, generally, American Indian Policy Review Commission, *Final Report.*

54. See L. R. Weatherhead, "What Is an 'Indian Tribe'?—The Question of Tribal Existence," *American Indian Law Review* 8, no. 1 (1980): 1–47; William W. Quinn, Jr., "Federal Acknowledgment of American Indian Tribes: The Historical Development of a Legal Concept," *American Journal of Legal History* 34 (October 1990): 331–364.

55. *Morton v. Mancari* 417 U.S. 535 (1974).

56. *Joint Tribal Council of the Passamaquoddy Tribe et al. v. Morton et al.* 528 F.2d 370 (1975).

57. Paul Brodeur, "Annals of Law," *New Yorker,* October 11, 1982, pp. 76–155; John M. R. Paterson and David Roseman, "A Reexamination of *Passamaquoddy v. Morton,*" *Maine Law Review* 31, no. 1 (1979): 115–151; Harry B. Wallace, "Indian Sovereignty and Eastern Indian Land Claims," *New York Law School Law Review* 27, no. 3 (1982): 921–950; Francis J. O'Toole and Thomas N. Tureen, "State Power and the Passamaquoddy Tribe: 'A Gross National Hypocrisy?'" *Maine Law Review* 23, no. 1 (1971): 1–39.

58. *State v. Newell* 84 Me. 465 (1892).

59. *Joint Tribal Council of the Passamaquoddy Tribe et al. v. Morton et al.* 528 F.2d 370 at 376–381 (1975).

60. U.S., *Statutes at Large* 94 (October 10, 1980): 1785–1795 [Maine Indian Claims Settlement Act].

61. See, for example, *Schaghticoke Tribe v. Kent School Corp.* 423 F. Supp. 780 (1976); *Mohegan Tribe v. Connecticut* 638 F.2d 612 (1980); *Narragansett Tribe v. Southern Rhode Island Land Development Corp.* 418 F. Supp. 798 (1976).

62. Quinn, "Federal Acknowledgment," p. 363. See also *Harjo v. Kleppe* 420 F. Supp. 1110 (1976); Sidney L. Harring, "Crazy Snake and the Creek Struggle for Sovereignty: The Native American Legal Culture and American Law," *American Journal of Legal History* 34 (October 1990): 365–380.

63. U.S., Department of the Interior, "Procedures for Establishing that an American Indian Group Exists as an Indian Tribe," *Code of Federal Regulations,* sec. 54, 25 (1980): 199–204; U.S., Department of the Interior, "Procedures for Establishing that an American Indian Group Exists as an Indian Tribe: Proposed Rule, Supplementary Information," *Federal Register* 43 (June 1, 1978): 23743–23746; U.S., Department of the Interior, "Indian Tribal Entities that Have a Government-to-Government Relationship with the United States," *Federal Register* 45 (April 24, 1980): 27828–27830.

64. Department of the Interior, "Procedures for Establishing an Indian Tribe," sec. 54, p. 200.

65. *United States v. Washington* 476 F. Supp. 1101 (1979).

66. *Mashpee Tribe v. New Seabury Corp. et al.* 592 F.2d 575 (1979). See also "The Unilateral Termination of Tribal Status: *Mashpee Tribe v. New Seabury Corp.,*" *Maine Law Review* 31, no. 1 (1979): 153–170.

67. Michael C. Walch, "Terminating the Indian Termination Policy," *Stanford Law Review* 35 (July 1983): 1188–1189; Department of the Interior, "Procedures for Establishing an Indian Tribe," sec. 54.7(g), p.202.

68. For a discussion of Menominee termination, see pp. 102–104.

69. U.S., *Statutes at Large* 87 (December 22, 1973): 770–773 [Menominee Restoration Act].

70. Ibid.; Walch, "Terminating Termination," p. 1193.

71. Beth Ritter Knoche, "Termination, Self-Determination and Restoration: The Northern Ponca Case" (M.A. thesis, University of Nebraska–Lincoln, 1990), p. 89.

72. Ibid., pp. 79–82, 98–99; Elizabeth S. Grobsmith and Beth R. Ritter, "The Ponca Tribe of Nebraska: The Process of Restoration of a Federally-Terminated Tribe," *Human Organization* 51 (Spring 1992): 1–16. See also Rennard Strickland and Charles F. Wilkinson, eds., *Felix S. Cohen's Handbook of Federal Indian Law* (Charlottesville, Va.: Michie Bobbs-Merrill, 1982), pp. 815–818.

73. Quoted in Grobsmith and Ritter, "Ponca Tribe of Nebraska," p. 1.

74. John R. White, "Barmecide Revisited: The Gratuitous Offset in Indian Claims Cases," *Ethnohistory* 25 (Spring 1978): 179.

75. Ibid., p. 183.

76. *Delaware Tribe of Indians v. United States* 21 Ind. Cl. Comm. 18 (1969).

77. White, "Barmecide Revisited," pp. 183–184.

78. Ibid., pp. 185–186; *Peoria Tribe of Indians of Oklahoma et al. v. United States* 390 U.S. 468 (1968).

79. White, "Barmecide Revisited," pp. 187–190. See also Edward Lazarus, *Black Hills/White Justice: The Sioux Nation Versus the United States, 1775 to the Present* (New York: HarperCollins, 1991).

80. U.S., *Statutes at Large* 88 (October 27, 1974): 1499–1500, sec. 2.

81. Russel Lawrence Barsh and James Youngblood Henderson, *The Road: Indian Tribes and Political Liberty* (Berkeley: University of California Press, 1980), pp. 94–95, 125, 256; Vine Deloria, Jr., and Clifford M. Lytle, *The Nations Within: The Past and Future of American Indian Sovereignty* (New York: Pantheon Books, 1984), p. 261.

82. U.S., *Statutes at Large* 82 (April 11, 1968): 77–81 [Indian Bill of Rights].

83. Carole E. Goldberg, "Public Law 280: The Limits of State Jurisdiction over Reservation Indians," *UCLA Law Review* 22 (February 1975): 550–551, 561; *Omaha Tribe v. Village of Walthill* 334 F. Supp. 823 (1971); *United States v. Brown* 334 F. Supp. 536 (1971).

Chapter 9

1. Duane Big Eagle, "Flowers of Winter: Four Songs," in *The Remembered Earth: An Anthology of Contemporary Native American Literature*, ed. Geary Hobson (Albuquerque: University of New Mexico Press, 1991), pp. 143–144.

2. Monroe E. Price, *Law and the American Indian: Readings, Notes, and Cases* (Indianapolis: Bobbs-Merrill, 1973).

3. "Treaty with the Nisqualli, Puyallup, Etc., 1854," in *Treaties and Agreements of the Indian Tribes of the Pacific Northwest* (Washington, D.C.: Institute for the Development of Indian Law, n.d.), p. 13.

4. U.S., *Statutes at Large* 67 (August 15, 1953): 589, sec. 2b [Public Law 280].

5. *Puyallup Tribe, Inc. v. Department of Game* 391 U.S. 392 (1968). This case is commonly referred to as Puyallup I. Subsequent litigation included *Puyallup Tribe, Inc. v. Department of Game* 414 U.S. 44 (1973), known as Puyallup II, and *Puyallup Tribe, Inc. v. Department of Game* 433 U.S. 165 (1977), known as Puyallup III.

6. Puyallup I, II, and III; Jack L. Landau, "Empty Victories: Indian Treaty Fishing Rights in the Pacific Northwest," *Environmental Law* 10 (Winter 1980): 414–437. Puyallup I was soundly criticized in the legal profession, and it is blamed for the massive amount of litigation that followed. See also *United States v. Washington* 476 F. Supp. 1101 (1979); James C. Giudici, "State Regulation of Indian Treaty Fishing Rights: Putting Puyallup III into Perspective," *Gonzaga Law Review* 13 (Fall 1977): 140–189; Sasha Harmon, "Writing History by Litigation: The Legacy and Limitations of Northwest Indian Rights Cases," *Columbia* 3 (Winter 1990–1991): 5–15.

7. *New Mexico v. Mescalero Apache Tribe* 462 U.S. 324 (1983). See also David H. Getches and Charles F. Wilkinson, *Federal Indian Law: Cases and Materials,* 2nd ed. (St. Paul: West, 1986), pp. 716–718.

8. *Kimball v. Callahan* 493 F.2d 564 (1974). See also Mary Pearson, "Hunting Rights: Retention of Treaty Rights After Termination—*Kimball v. Callahan*," *American Indian Law Review* 4, no. 1 (1976): 121–133.

9. *Montana v. United States* 450 U.S. 544 (1981); S. J. Bloxham, "Tribal Sovereignty: An Analysis of *Monta[na] v. United States*," *American Indian Law Review* 8, no. 1 (1980): 175–181.

10. U.S., *Statutes at Large* 87 (December 28, 1973): 884–903 [Endangered Species Act].

11. U.S., *Statutes at Large* 54 (June 8, 1940): 250–251 [Eagle Protection Act].

12. U.S., *Statutes at Large* 76 (October 24, 1962): 1246.

13. *United States v. Dion* 476 U.S. 734 (1986).

14. Lloyd Burton, *American Indian Water Rights and the Limits of Law: Reflections in a Glass Bead* (Lawrence: University Press of Kansas, 1991), p. 31.

15. Ibid., pp. 35–62. See also William Douglas Back and Jeffrey S. Saylor, "Navajo Water Rights: Pulling the Plug on the Colorado River?" *Natural Resources Journal* 20 (January 1980): 71–90; Robert D. Dellwo, "Recent Developments in the Northwest Regarding Indian Water Rights," *Natural Re-*

sources *Journal* 20 (January 1980): 101–120; Gwendolyn Griffith, "Indian Claims to Groundwater: Reserved Rights or Beneficial Interest?" *Stanford Law Review* 33 (November 1980): 103–130; Robert S. Pelcyger, "The *Winters* Doctrine and the Greening of the Reservation," *Journal of Contemporary Law* 4 (Winter 1977): 19–37; Robert Isham, Jr., *"Colville Confederated Tribes v. Walton:* Indian Water Rights and Regulation in the Ninth Circuit," *Montana Law Review* 43 (Spring 1982): 247–269; Harold A. Ranquist, "The *Winters* Doctrine and How It Grew: Federal Reservation of Rights to the Use of Water," *Brigham Young University Law Review,* no. 4 (1975): 639–724; Kenneth E. Foster, "The *Winters* Doctrine: Historical Perspective and Future Applications of Reserved Water Rights in Arizona," *Ground Water* 16 (May–June 1978): 186–188.

16. *Cappaert v. United States* 426 U.S. 128 (1976).

17. *In re General Adjudication of All Rights to Use Water in the Big Horn River System* 753 P.2d 76 (1988).

18. *Shoshone Tribe v. Wyoming* 753 P.2d 76 (1988), *cert. denied* 57 U.S.L.W. 3860 (1989); *Wyoming v. United States* 488 U.S. 1040 (1989). For a discussion of these confusing and alarming cases, see Burton, *American Indian Water Rights,* pp. 38–40.

19. Burton, *American Indian Water Rights,* pp. 69–86.

20. Marjane Ambler, *Breaking the Iron Bonds: Indian Control of Energy Development* (Lawrence: University Press of Kansas, 1990), p. 29; Donald L. Fixico, "Tribal Leaders and the Demand for Natural Energy Resources on Reservation Lands," in *The Plains Indians of the Twentieth Century,* ed. Peter Iverson (Norman: University of Oklahoma Press, 1985), p. 220.

21. Quoted in Fixico, "Energy Resources on Reservation Lands," p. 222.

22. Ambler, *Breaking the Iron Bonds,* pp. 91–96.

23. U.S., *Statutes at Large* 96 (December 22, 1982): 1938–1940 [Indian Mineral Development Act].

24. U.S., *Statutes at Large* 96 (January 7, 1983): 2201–2263 [Nuclear Waste Policy Act].

25. Nancy E. Hovis, "Tribal Involvement Under the Nuclear Waste Policy Act of 1982: Education by Participation," *Journal of Environmental Law and Litigation* 3 (1988): 45–65.

26. *Brendale v. Confederated Tribes and Bands of the Yakima Indian Nation* 492 U.S. 408 (1989).

27. *United States v. Mazurie* 419 U.S. 544 (1975).

28. Jessica S. Gerrard, "Undermining Tribal Land Use Regulatory Authority: *Brendale v. Confederated Tribes,"* *University of Puget Sound Law Review* 13 (Winter 1990): 349–375.

29. Daniel H. Israel, "The Reemergence of Tribal Nationalism and Its Im-

pact on Reservation Resource Development," *University of Colorado Law Review* 47 (Summer 1976): 617–652.

30. Russel Lawrence Barsh, "Issues in Federal, State, and Tribal Taxation of Reservation Wealth: A Survey and Economic Critique," *Washington Law Review* 54 (June 1979): 537–542.

31. *McClanahan v. State Tax Commission of Arizona* 411 U.S. 164 (1973); Russell W. Davisson, "Indian Law—Taxation—Reservation Indian's Income Not Taxable If Derived from Reservation Sources—State Power over Reservation Indians Is Limited," *University of Kansas Law Review* 22 (Spring 1974): 471–479. For the history of taxation as applied to the Sioux, see James R. McCurdy, "Federal Income Taxation and the Great Sioux Nation," *South Dakota Law Review* 22 (Spring 1977): 296–321.

32. Carole E. Goldberg, "A Dynamic View of Tribal Jurisdiction to Tax Non-Indians," *Law and Contemporary Problems* 40 (Winter 1976): 166–189.

33. *Moe v. Confederated Salish & Kootenai Tribes of the Flathead Reservation* 425 U.S. 463 (1976). For background on Chief Justice William Rehnquist's anti-Indian bias in this case, see Bob Woodward and Scott Armstrong, *The Brethren: Inside the Supreme Court* (New York: Avon, 1979), p. 490.

34. *Merrion v. Jicarilla Apache Tribe* 455 U.S. 130 (1982).

35. *Kerr-McGee Corporation v. Navajo Tribe* 471 U.S. 195 (1985).

36. *Cotton Petroleum Corp. v. New Mexico* 490 U.S. 163 (1989).

37. Ambler, *Breaking the Iron Bonds,* pp. 197–201; Samuel Fabbraio, Jr., "Tribal Severance Taxes: The Uncertain Sovereign Function: *Merrion v. Jicarilla Apache Tribe,*" *University of Bridgeport Law Review* 4 (1982): 133–151; Katherine B. Crawford, "State Authority to Tax Non-Indian Oil & Gas Production on Reservations: *Cotton Petroleum Corp. v. New Mexico,*" *Utah Law Review,* no. 2 (1989): 495–519; "BIA Supports Wind River Tax," *American Indian Report* 4 (May 1988): 4; "Senecas and State Reach Tentative Tax Agreement," *American Indian Report* 5 (February 1989): 4.

38. Linda A. Marousek, "The Indian Child Welfare Act of 1978: Provisions and Policy," *South Dakota Law Review* 25 (Winter 1980): 99 n.11.

39. William Byler, "The Destruction of American Indian Families," in *The Destruction of American Indian Families,* ed. Steven Unger (New York: Association of American Indian Affairs, 1977), pp. 1–5.

40. Aileen Red Bird and Patrick Melendy, "Indian Child Welfare in Oregon," in Unger, ed., *Destruction of American Indian Families,* p. 45.

41. David Woodward, "The Rights of Reservation Parents and Children: Cultural Survival or the Final Termination?" *American Indian Law Review* 3, no. 1 (1975): 22.

42. Jane G. Printz, "Navajo Grandparents—'Parent' or 'Stranger'—A Child Custody Determination," *New Mexico Law Review* 9 (Winter 1978–1979): 187–194; Gaylene J. McCartney, "The American Indian Child–Welfare Crisis: Cul-

tural Genocide or First Amendment Preservation?'' *Columbia Human Rights Law Review* 7 (Fall–Winter 1975–1976): 529–551.

43. *Potawatomies of the Hannaville Indian Community v. Houston* 393 F. Supp. 719 (1973). See also McCartney, ''Indian Child–Welfare Crisis,'' pp. 534–535.

44. Woodward, ''Rights of Reservation Parents and Children,'' pp. 21, 32, 40–42.

45. U.S., *Statutes at Large* 92 (November 8, 1978): 3069–3078 [Indian Child Welfare Act].

46. Ibid., p. 3069.

47. Ibid., Title I, sec. 101a–d, p. 3071; sec. 102e–f, p. 3072.

48. Ibid., sec. 105b, p. 3073; sec. 102a, p. 3071.

49. Ibid., sec. 102b, p. 3073; sec. 108a–c, p. 3074; Titles II and III, pp. 3075–3077.

50. Joan Heifetz Hollinger, ''Beyond the Best Interests of the Tribe: The Indian Child Welfare Act and the Adoption of Indian Children,'' *University of Detroit Law Review* 66 (Spring, 1989): 451–501; Manuel P. Guerrero, ''Indian Child Welfare Act of 1978: A Response to the Threat to Indian Culture Caused by Foster and Adoptive Placements of Indian Children,'' *American Indian Law Review* 7, no. 1 (1979): 51–77; Marousek, ''Indian Child Welfare Act,'' pp. 105–114. See also Garry Wamser, ''Child Welfare Under the Indian Child Welfare Act of 1978: A New Mexico Focus,'' *New Mexico Law Review* 10 (Summer 1980): 413–429; Russel Lawrence Barsh, ''The Indian Child Welfare Act of 1978: A Critical Analysis,'' *Hastings Law Journal* 31 (July 1980): 1287–1336. The Supreme Court upheld the constitutionality of the ICWA in *Mississippi Band of Choctaw Indians v. Holyfield et al.* 490 U.S. 30 (1989).

51. Quoted in U.S. Congress, Senate, *American Indian Religious Freedom: Hearings on S.J. Res. 102 Before the Senate Select Committee on Indian Affairs,* 95th Cong., 2nd sess., 1978, p. 193.

52. Ibid., pp. 86–87, quoted in Howard Stambor, ''Manifest Destiny and American Indian Religious Freedom: *Sequoyah, Badoni,* and the Drowned Gods,'' *American Indian Law Review* 10, no. 1 (1982): 59.

53. U.S., *Statutes at Large* 92 (August 11, 1978): 469–470 [American Indian Religious Freedom Act].

54. Ibid., p. 469.

55. Ibid., p. 470.

56. T. J. Ferguson, ''The American Indian Religious Freedom Act and Zuni Pueblo'' (Paper presented at the annual meeting of the American Society for Ethnohistory, Colorado Springs, Colorado, November 1981); Dean B. Suagee, ''American Indian Religious Freedom and Cultural Resources Management: Protecting Mother Earth's Caretakers,'' *American Indian Law Review* 10, no. 1 (1982): 1–58.

57. Kathryn Harris, "The American Indian Religious Freedom Act and Its Promise," *American Indian Journal* 5 (June 1979): 7–10; Jill E. Martin, "Constitutional Rights and Indian Rites: An Uneasy Balance," *Western Legal History* 3 (Fall–Winter 1990): 245–270.

58. *Sequoyah v. Tennessee Valley Authority* 620 F.2d 1159 (1980).

59. *Badoni v. Higginson* 638 F.2d 172 (1980). See also *Crow v. Gullet* 706 F.2d 856 (1983); *Inupiat Community v. Alaska* 746 F.2d 570 (1984); *Wilson v. Block* 708 F.2d 735 (1983); Donnal S. Mixon, "Native American Site Specific Free Exercise Claims to Public Land" (manuscript, Texas Tech University School of Law, 1983), pp. 1–36; Randolf J. Rice, "Native Americans and the Free Exercise Clause," *Hastings Law Journal* 28 (July 1988): 1509–1536; Laurie Ensworth, "Native American Free Exercise Rights to the Use of Public Lands," *Boston University Law Review* 63, no. 1 (1983): 141–179. See also *Frank v. Alaska* 604 P.2d 1068 (1979); *"Frank v. Alaska:* An Athabascan's Right to Practice His Religion," *Americans Before Columbus,* National Indian Youth Council, special ed., April 1982, pp. 6, 10.

60. *Lyng v. Northwest Indian Cemetery Protective Ass'n* 485 U.S. 439 (1988).

61. Stephen McAndrew, *"Lyng v. Northwest Indian Cemetery Protective Ass'n:* Closing the Door to Indian Religious Sites," *Southwestern University Law Review* 18, no. 4 (1989): 603–629; Charles Miller, "The Navajo–Hopi Relocation Act and the First Amendment Free Exercise Clause," *University of San Francisco Law Review* 23 (Fall 1988): 97–121.

62. *Employment Division, Department of Human Resources of Oregon, et al. v. Alfred L. Smith et al.* 494 U.S. 872 (1990): Scalia majority opinion, pp. 872–890; Blackmun dissent, pp. 907–921; O'Connor concurring opinion, pp. 890–906.

63. Ibid., Scalia opinion, p. 877.

Chapter 10

1. Of the many publications that appeared, very few explained the nature of Indians and the Bill of Rights. Among the myths frequently mentioned were that the Bill of Rights applied to *all* individuals in the United States; that it protected *all* their religious, political, and economic freedoms; and that it guaranteed equal treatment for *all* people. See, for example, ". . . do ordain and establish this Constitution for the United States of America," *The Bicentennial of the Bill of Rights,* vol. 18, comp. Project !87, American Historical Association and the American Political Science Association (Richmond, Va.: Byrd Press, 1991), particularly Herman Belz, "A Chronology of Civil Liberties and Civil Rights in the United States," front cover. See also Ellen Alderman and Caroline Kennedy, *In Our Defense: The Bill of Rights in Action* (New York: Morrow, 1991).

2. Elizabeth Cook-Lynn, *From the River's Edge* (Boston: Little, Brown, 1991), p. 126.

3. Quoted in Hank Whittemore, "She Leads a Nation," *Parade,* August 18, 1991, pp. 4–5.

4. Quoted in Michael Satchell, "These Battles the Indians Are Winning," *Parade,* June 17, 1979, pp. 9, 11. See also *The NARF Legal Review* (formerly *Announcements*), the newsletter of the Native American Rights Fund.

5. *Seminole Tribe of Florida v. Butterworth* 658 F.2d 310 (1981); *Barona Group of the Capital Grande Band of Mission Indians v. Duffy* 694 F.2d 1185 (1982); *United States v. Farris* 624 F.2d 890 (1980); *United States v. Dakota* 796 F.2d 186 (1986).

6. Wayne Beissert, "Games Boost Economies," *USA Today,* December 15, 1989, p. 8A.

7. *Cabazon Band of Mission Indians et al. v. California* 480 U.S. 202 (1987).

8. Allen C. Turner, "Evolution, Assimilation, and State Control of Gambling in Indian Country: Is *Cabazon v. California* an Assimilationist Wolf in Preemption Clothing?" *Idaho Law Review* 24, no. 2 (1987–1988): 332–335.

9. U.S., *Statutes at Large* 102 (October 17, 1988): 2467–2488 [Indian Gaming Regulatory Act].

10. Ibid., sec. 4, pp. 2467–2469.

11. Ibid., sec. 7, pp. 2470–2471; sec. 11, pp. 2472–2479.

12. Ibid., sec. 11, pp. 2477–2479; sec. 20, pp. 2485–2486.

13. "State Joins Wisconsin Against Indian Ruling," *Lincoln* [Nebraska] *Star,* September 27, 1991; "Connecticut Indian Casino Rocks Nation's Gaming Boat," *Lincoln* [Nebraska] *Journal-Star,* May 26, 1991; Lindsey Gruson, "U.S. Approves Indian Casino Plan in Connecticut," *New York Times,* May 26, 1991, p. 5; Andrea Stone and Mark Mayfield, "Off-Reservation Gambling," *USA Today,* July 31, 1990; Margaret Nelson, " 'Caesars' in Minnesota," *USA Today,* July 31, 1990; John Larrabee, "Connecticut Tribe Is on a Roll," *USA Today,* May 9, 1991.

14. Gary Newman and Paul Goodsell, "Lujan Ruling Protects Iowa Tribes," *Omaha World-Herald,* December 19, 1991; David Thompson, "Lujan Skirts Key Issue in Rejecting Casino," *Omaha World-Herald,* December 19, 1991; Victoria Benning, "2 Nebraska Tribes Near Casino Pacts with Iowa," *Omaha World-Herald,* December 19, 1991; Joe Brennan, "Missouri Flows Through Middle of Casino Clash," *Omaha World-Herald,* June 16, 1991; "Policy on Buying Indian Land Should Apply to Casino Plan" (editorial), *Omaha World-Herald,* August 4, 1991; "Winnebago Tribe Near Agreement on Casino," *Lincoln Star,* December 12, 1991; "Winnebagoes Oppose Council Bluffs Casino," *Lincoln Star,* December 17, 1991; "Omaha Tribe Signs Contract for Gambling," *Lincoln Star,* September 9, 1991; "Injunction Puts Winnebago Tribe's Sloan Casino Plans on Hold," *Lincoln Star,* July 5, 1991; "Interior Secretary

Scuttles Plans for Sioux Casino,'' *Lincoln Star,* December 19, 1991; ''Nebraskans Applaud Lujan's decision,'' *Lincoln Star,* December 19, 1991; ''Omaha Tribe Allowed to Build Iowa Casino,'' *Lincoln Star,* December 31, 1991.

15. ''Developers Twisting Indian Sovereignty into Exploitation,'' *Omaha World-Herald,* September 8, 1991; ''Heat, Scandal Close Indian Bingo Hall,'' *Omaha World-Herald,* September 8, 1991; ''Small Tribe Hits It Big with Casino,'' *Omaha World-Herald,* July 21, 1991; Paul Hammel, ''A Prairie Las Vegas . . . ,'' *Omaha World-Herald,* August 4, 1991.

16. Jack F. Trope and Walter R. Echo-Hawk, ''The Native American Graves Protection and Repatriation Act: Background and Legislative History'' (Manuscript prepared for the Native American Rights Fund and the Association on American Indian Affairs, 1991), pp. 1–7.

17. Quoted in Robert Bieder, ''A Brief Historical Survey of the Expropriation of American Indian Remains,'' in U.S. Congress, Senate, 101st Cong., 2nd sess., May 14, 1990, pp. 278–363, quoted in ibid., p. 8.

18. *Wanna the Bear v. Community Construction* 128 Cal. App. 3d 536 (1982); *State v. Glass* 273 N.E. 2d 893 (1971).

19. Trope and Echo-Hawk, ''Native American Graves Protection and Repatriation Act,'' pp. 14–15. For a list of states, see p. 54, n.41.

20. Ibid., pp. 22–23. For the long, continuing struggle over repatriation in Nebraska, see Orlan J. Svingen, ''The Pawnee of Nebraska: Twice Removed,'' and Roger C. Echo-Hawk and Walter R. Echo-Hawk, ''Battlefields and Burial Grounds: The Indian Struggle to Protect Ancestral Graves and Human Remains in the United States'' (Both papers delivered at the annual meeting of the Western History Association, Austin, Texas, October 15, 1991); Orlan J. Svingen, ''History of the Expropriation of Pawnee Indian Graves in the Control of the Nebraska State Historical Society'' (Manuscript prepared for the Native American Rights Fund, January 25, 1989); ''Some Questions and Answers About the Human Remains Issue,'' *Newsletter of the Nebraska State Historical Society* 41 (February 1989): 1–4; ''Genoa: Tribe Must Pay Bills Before It Makes More Burials,'' *Lincoln Star,* August 16, 1991; ''Pawnee File Formal Grievance,'' *Lincoln Star,* August 1, 1990; Betty Stevens, ''Burial of Ancestors a Bittersweet Event,'' *Lincoln Star,* October 14, 1991; Robynn Tysver, ''Historical Society Gives Up Contested Indian Burial Goods,'' *Lincoln Star,* September 10, 1991; ''Indian Burial Sites Said Still Protected,'' *Lincoln Journal-Star,* August 10, 1991; Margaret Reist, ''Pawnees Hail Ruling Backing Right to Historical Society Records,'' *Lincoln Journal-Star,* June 1, 1991; ''Pawnees to Get George III Medal Back,'' *Lincoln Journal-Star,* November 2, 1991; Wendy Mott, ''Itskari Indian Remains Finally Laid to Rest,'' *Daily Nebraskan* [Lincoln], September 30, 1991; ''Society to Give More Remains to Pawnees,'' *Omaha World-Herald,* June 16, 1991; Orlan J. Svingen, ed., ''History of LB 340'' (manuscript, 1992).

The return of cultural patrimony also began in 1990 in Nebraska with the completion of negotiations between the Omaha Tribal Nation and the Peabody Museum of Harvard University. The Sacred Pole, white buffalo robe, pipe, and other artifacts were restored to the Omahas with the assistance of the Center for Great Plains Studies and the State Museum of the University of Nebraska–Lincoln (Cindy Connolly, "Omaha Tribe Celebrates Return of Sacred Artifacts," *Omaha World-Herald,* August 5, 1991).

21. U.S., *Statutes at Large* 104 (November 16, 1990): 3048–3058 [Native American Graves Protection and Repatriation Act].

22. Quotation from *Congressional Record,* 101st Cong., 1st sess., 1990, p. 17174, quoted in Trope and Echo-Hawk, "Native American Graves Protection and Repatriation Act," p. 29. See also Felicity Barringer, "Bush Weighs Signing of Law on Indian Artifacts," *New York Times,* November 4, 1990, p. 25; Deward E. Walker, Jr., "Anthropologists Must Allow American Indians to Bury Their Dead," *Chronicle of Higher Education,* September 12, 1990, p. B2.

23. U.S., Native American Graves Protection and Repatriation Act, pp. 3048–3058. See also Jane Hubert, "A Proper Place for the Dead: A Critical Review of the 'Reburial' Issue," *Journal of Indigenous Studies* 1 (Winter 1989): 28–62; Lawrence Rosen, "The Excavation of American Indian Burial Sites: A Problem in Law and Professional Responsibility," *American Anthropologist* 82 (March 1980): 5–27; Bowen Blair, "Indian Rights: Native Americans Versus American Museums—A Battle for Artifacts," *American Indian Law Review* 7, no. 1 (1979): 125–154.

24. Karl N. Llewellyn and E. Adamson Hoebel, *The Cheyenne Way: Conflict and Case Law in Primitive Jurisprudence* (Norman: University of Oklahoma Press, 1941), pp. 9–12.

25. Ibid., p. 10.

26. Ibid., pp. 10–11.

27. Ibid., p. 11.

28. *Powwow Highway,* Hand Made Films, prod. Jan Wieringo, dir. Jonathan Wacks, screenplay adapted from *Powwow Highway,* by David Seals, 1989 (91 min.).

29. Ibid.

Bibliographical Essay

Writing Native American legal history requires placing legal and constitutional history and Indian history into a common framework, and it necessarily depends on primary and secondary materials from several subdisciplines that have not frequently been combined. Nevertheless, excellent materials are available for writing and reading about Native American legal history.

On the legal side, invaluable original source materials can be found in *United States Reports,* the various federal district and appellate court reports, and the state supreme court reports. The number of cases in Indian law has accelerated significantly since World War II. Because Indians have been the subject of so much legislation, those interested in this topic should also turn to United States *Statutes at Large* and to the records of the corresponding congressional debates. Numerous acts were passed in both the nineteenth and twentieth centuries.

Secondary law materials are abundant in law reviews and law journals available in most law libraries. Most essays concern a particular court case, a particular statute, or a particular aspect of legal doctrine. Overview articles are rare. Readers should begin with the *American Indian Law Review,* published at the University of Oklahoma. Other law reviews contain comments and feature articles about Indian law relevant to their state. Additional articles are available in history and anthropology scholarly journals, especially *Ethnohistory, American Indian Culture and Research Journal, American Indian Quarterly, Journal of American History,* and various regional and state journals.

Bibliographies and reference materials are indispensable. Particularly helpful bibliographies include Francis Jennings, gen. ed., and William R. Swagerty, asst. ed., *The Newberry Library Center for the History of the American Indian Bibliographical Series* (Bloomington: Indiana University Press, 1978–   ); and Francis Paul Prucha, *A Bibliographical Guide to the History of Indian–White Relations in the United States* (Chicago: University of Chicago Press, 1977), covering publications up to 1975, and *Indian–White Relations in the United States: A Bibliography of Works Published, 1975–1980* (Lincoln: University of Nebraska Press, 1982), an additional five-year supplement. Useful reference guides to Indian case law and statutory law include Rennard Strickland and Charles F. Wilkinson, eds., *Felix S. Cohen's Handbook of Federal Indian Law* (Charlottesville, Va.: Michie Bobbs-Merrill, 1982); Imre Sutton, *Indian Land Tenure: Bibliographical Essays and a Guide to the Literature* (New York: Clearwater, 1975); and William C. Canby, Jr., *American Indian Law in a Nutshell* (St. Paul: West, 1981).

The first text in Indian law was written in 1973. That text, Monroe E. Price, *Law and the American Indian: Readings, Notes, and Cases* (Indianapolis: Bobbs-Merrill, 1973), still competes with David H. Getches and Charles F. Wilkinson, *Federal Indian Law: Cases and Materials,* 2nd ed. (St. Paul: West, 1986). Overview treatments for particular time periods or aspects of Indian legal history include Vine Deloria, Jr., and Clifford M. Lytle, *American Indians, American Justice* (Austin: University of Texas Press, 1983); Charles N. Wilkinson, *American Indians, Time, and the Law: Native Societies in a Modern Constitutional Democracy* (New Haven, Conn.: Yale University Press, 1987); Vine Deloria, Jr., and Clifford Lytle, *The Nations Within: The Past and Future of American Indian Sovereignty* (New York: Pantheon Books, 1984); Russel Lawrence Barsh and James Youngblood Henderson, *The Road: Indian Tribes and Political Liberty* (Berkeley: University of California Press, 1980); Petra T. Shattuck and Jill Norgren, *Partial Justice: Federal Indian Law in a Liberal Constitutional System* (New York: Berg, 1991); and Imre Sutton, ed., *Irredeemable America: The Indians' Estate and Land Claims* (Albuquerque: University of New Mexico Press, 1985).

Texts in Indian history that offer insight into political and legal dimensions include Francis Paul Prucha, *The Great Father: The United States Government and American Indians,* 2 vols. (Lincoln: University of Nebraska Press, 1984), a definitive treatment of federal policy toward

Native Americans; Wilcomb E. Washburn, *Red Man's Land/White Man's Law: A Study of the Past and Present Status of the American Indian* (New York: Scribner, 1971), a very readable but dated survey; Arrell Morgan Gibson, *The American Indian: Prehistory to the Present* (Lexington, Mass.: Heath, 1980), a comprehensive text running from the origins of Native Americans to the present, with significant emphasis on the twentieth century; Angie Debo, *A History of the Indians of the United States* (Norman: University of Oklahoma Press, 1970); and William T. Hagan, *American Indians,* rev. ed. (Chicago: University of Chicago Press, 1979).

Early Native American history, what anthropologists and archaeologists call "prehistory," is treated thoroughly in Peter Farb, *Man's Rise to Civilization: The Cultural Ascent of the Indians of North America* (New York: Dutton, 1968); Dean Snow, *The American Indians: Their Archaeology and Prehistory* (London: Thames and Hudson, 1976); and Edward H. Spicer, *A Short History of the Indians of the United States* (New York: Van Nostrand Reinhold, 1969).

Collections of essays covering entire time periods or specific regions or specific Native American leaders are useful. Recent contributions include Roger Nichols, ed., *The American Indian: Past and Present,* 4th ed. (New York: McGraw-Hill, 1992); Peter Iverson, ed., *The Plains Indians of the Twentieth Century* (Norman: University of Oklahoma Press, 1985); Jane F. Smith and Robert M. Kvasnicka, eds., *Indian–White Relations: A Persistent Paradox* (Washington, D.C.: Howard University Press, 1976); James S. Olson and Raymond Wilson, *Native Americans in the Twentieth Century* (Urbana: University of Illinois Press, 1984); L. G. Moses and Raymond Wilson, eds., *Indian Lives: Essays on Nineteenth- and Twentieth-Century Native American Leaders* (Albuquerque: University of New Mexico Press, 1985); and R. David Edmunds, ed., *American Indian Leaders: Studies in Diversity* (Lincoln: University of Nebraska Press, 1980).

Rights of Indigenous Peoples

Anthropologists have contributed the most to writings on the rights of indigenous peoples, although some philosophers and legal scholars also have addressed this subject.

For an introduction to legal anthropology, see Ted C. Lewellen, *Political Anthropology: An Introduction* (Westport, Conn.: Bergin and

Garvey, 1983), and June Starr and Jane F. Collier, eds., *History and Power in the Study of Law: New Directions in Legal Anthropology* (Ithaca, N.Y.: Cornell University Press, 1989). Of particular significance are the works of E. Adamson Hoebel, especially *The Law of Primitive Man: A Study in Comparative Legal Dynamics* (Cambridge, Mass.: Harvard University Press, 1954); E. Adamson Hoebel and Karl N. Llewellyn, *The Cheyenne Way: Conflict and Case Law in Primitive Jurisprudence* (Norman: University of Oklahoma Press, 1941); and E. Adamson Hoebel and Ernest Wallace, *The Comanches: Lords of the South Plains* (Norman: University of Oklahoma Press, 1952). To understand the concepts of rights from a philosophical perspective, see Judith Jarvis Thomson, *The Realm of Rights* (Cambridge, Mass.: Harvard University Press, 1990). A good ethnohistorical survey is D'Arcy McNickle, *They Came Here First: The Epic of the American Indian* (Philadelphia: Lippincott, 1949), particularly pp. 75–91.

From a specific tribal viewpoint, on the Cheyenne, see Hoebel and Llewellyn, *Cheyenne Way;* and Robert A. Fairbanks, "The Cheyenne and Their Law: A Positivist Inquiry," *Arkansas Law Review* 32 (Fall 1978): 403–445, and "A Discussion of the Nation-State Status of American Indian Tribes: A Case Study of the Cheyenne Nation," *American Indian Journal* 3 (October 1977): 2–24. On the Senecas, see Anthony F. C. Wallace, *The Death and Rebirth of the Seneca* (New York: Vintage Books, 1972). On the Cherokees, see John Phillip Reid, *A Law of Blood: The Primitive Law of the Cherokee Nation* (New York: New York University Press, 1970), and *A Better Kind of Hatchet: Law, Trade, and Diplomacy in the Cherokee Nation During the Early Years of European Contact* (University Park: Pennsylvania State University Press, 1976); Rennard Strickland, *Fire and the Spirits: Cherokee Laws from Clan to Court* (Norman: University of Oklahoma Press, 1975); and John R. Wunder, "'Doesn't Anyone Speak *Injun* in This Courtroom?': New Perspectives in Native American Legal History," *Reviews in American History* 5 (December 1977): 467–476. On the Kiowas, see John R. Wunder, *The Kiowa* (New York: Chelsea House, 1989), and Jane Richardson, *Law and Status Among the Kiowa Indians* (New York: American Ethnological Society, 1940). And on the Creeks, see Sidney L. Harring, "Crazy Snake and the Creek Struggle for Sovereignty: The Native American Legal Culture and American Law," *American Journal of Legal History* 34 (October 1990): 365–380.

The Colonial Era and the Nineteenth Century

Much has been written about the history of Native Americans in the colonial era and the nineteenth century. Two excellent books offer provocative discussions concerning European legal concepts of Indians before European settlement in North and South America and during the colonial era: Robert A. Williams, Jr., *The American Indian in Western Legal Thought: The Discourses of Conquest* (New York: Oxford University Press, 1990), and L. C. Green and Olive P. Dickason, *The Law of Nations and the New World* (Edmonton: University of Alberta Press, 1989).

Native American influence on the American Revolution and the Constitution has been the subject of much recent debate. See Donald A. Grinde, Jr., *The Iroquois and the Founding of the American Nation* (San Francisco: Indian Historian Press, 1977), and Bruce E. Johansen, *Forgotten Founders: Benjamin Franklin, the Iroquois, and the Rationale for the American Revolution* (Ipswich, Mass.: Gambit, 1982).

The nineteenth century up to 1871 was an uncertain time in the legal relationships of American Indians within the United States constitutional framework. A variety of topics growing out of specific laws, incidents, or federal government–Indian relationships have been investigated. For a discussion of dependency theory and its application to three Indian nations, see Richard White, *Roots of Dependency: Subsistence, Environment, and Social Change Among the Choctaws, Pawnees, and Navajos* (Lincoln: University of Nebraska Press, 1983). The Cherokee cases have been treated in the Indian history surveys noted earlier. An excellent analysis can be found in Joseph C. Burke, "The Cherokee Cases: A Study in Law, Politics, and Morality," *Stanford Law Review* 21 (February 1969): 500–531. The end of treaty making via the Resolution of 1871 is explored in John R. Wunder, "No More Treaties: The Resolution of 1871 and the Alteration of Indian Rights to Their Homelands," in *Working the Range: Essays on the History of Western Land Management and the Environment,* ed. John R. Wunder (Westport, Conn.: Greenwood Press, 1985), pp. 39–56.

After 1871, a number of important events influenced the direction of Native American history. The seminal case of *Lone Wolf v. Hitchcock* is treated in Ann Laquer Estin, *"Lone Wolf v. Hitchcock:* The Long Shadow," in *The Aggressions of Civilization: Federal Indian Policy Since the 1880s,* ed. Sandra L. Cadwalader and Vine Deloria, Jr. (Philadelphia: Temple University Press, 1984), pp. 215–245. Similarly, for *Ex parte Crow Dog,* see Sidney L. Harring, "Crow Dog's Case: A Chapter in the Legal History of Tribal Sovereignty," *American Indian Law Review* 14, no. 2 (1989): 191–239. The legal ramifications of assimilation are discussed in Frederick E. Hoxie, *A Final Promise: The Campaign to Assimilate the Indians, 1880–1920* (Lincoln: University of Nebraska Press, 1984), and Francis Paul Prucha, ed., *Americanizing the American Indians: Writings by the "Friends of the Indian," 1880–1990* (Lincoln: University of Nebraska Press, 1978).

The Twentieth Century

The twentieth century has seen a massive outpouring of materials relating to Native American legal history. Again, as in the previously discussed historiography, much of the discourse is topical.

Water law has received a great deal of comment. See Harold A. Ranquist, "The *Winters* Doctrine and How It Grew: Federal Reservation of Rights to the Use of Water," *Brigham Young University Law Review,* no. 4 (1975): 639–724; Lloyd Burton, *American Indian Water Rights and the Limits of Law: Reflections in a Glass Bead* (Lawrence: University Press of Kansas, 1991); Robert S. Pelcyger, "The *Winters* Doctrine and the Greening of the Reservation," *Journal of Contemporary Law* 4 (Winter 1977): 19–37; Norris Hundley, Jr., "The Dark and Bloody Ground of Indian Water Rights: Confusion Elevated to Principle," *Western Historical Quarterly* 9 (October 1978): 455–482; and Gwendolyn Griffith, "Indian Claims to Groundwater: Reserved Rights or Beneficial Interest?" *Stanford Law Review* 33 (November 1980): 103–130.

Fishing and hunting rights are discussed in Jack L. Landau, "Empty Victories: Indian Treaty Fishing Rights in the Pacific Northwest," *Environmental Law* 10 (Winter 1980): 413–456; James C. Giudici, "State Regulation of Indian Treaty Fishing Rights: Putting Puyallup III into Perspective," *Gonzaga Law Review* 13 (Fall 1977): 140–189; John C. Garland, *"Sohappy v. Smith:* Eight Years of Litigation over Indian Fishing Rights," *Oregon Law Review* 56 (1977): 680–701; and Mary Pearson, "Hunting Rights: Retention of Treaty Rights After Termination—*Kimball v. Callahan,"* *American Indian Law Review* 4, no. 1 (1976): 121–133.

The Indian New Deal has been the subject of a significant historiography. The Depression and Native Americans, the Indian Reorganization Act, the evolution of constitutions and tribal governments, voting rights, World War II, and the Indian Claims Commission are among the important topics considered. Kenneth R. Philp, *John Collier's Crusade for Indian Reform, 1920–1954* (Tucson: University of Arizona Press, 1977), is an excellent survey, and William H. Kelly, ed., *Indian Affairs and the Indian Reorganization Act: The Twenty Year Record* (Tucson: University of Arizona, 1954), is a collection of important essays. Among the many analyses, see Lawrence C. Kelly, "The Indian Reorganization Act: The Dream and the Reality," *Pacific Historical Review* 44 (August 1975): 291–312, and "John Collier and the Indian New Deal: An Assessment," in *Indian–White Relations: A Persistent Paradox,* ed. Jane F. Smith and Robert M. Kvasnicka (Washington, D.C.: Howard University Press, 1976), pp. 227–241; Theodore H. Haas, *Ten Years of Tribal Government Under I.R.A.,* Tribal Relations Pamphlets, no. 1 (Chicago: United States Indian Service, 1947); Roger Bromert, "The Sioux and the Indian New Deal, 1933–1946" (Ph.D. diss., University of Toledo, 1980), and "The Sioux and the Indian-CCC," *South Dakota History* 8 (Fall 1978): 340–356; Donald L. Parman, "The Indian and the Civilian Conservation Corps," *Pacific Historical Review* 40 (February 1971): 39–57; Peter M. Wright, "John Collier and the Oklahoma Indian Welfare Act of 1936," *Chronicles of Oklahoma* 50 (Autumn 1972): 347–371; Frank Ducheneaux, "The Indian Reorganization Act and the Cheyenne River Sioux," *American Indian Journal* 2 (August 1976): 8–14; Lynn Kickingbird, "Attitudes Toward the Indian Reorganization Bill," *American Indian Journal* 2 (July 1976): 8–14; Robert W. Oliver, "The Legal Status of American Indian Tribes," *Oregon Law Review* 38 (April 1959): 193–245; Jerry L. Bean, "The Limits of Indian Tribal Sovereignty: The Cornucopia of Inherent Powers," *North Dakota Law Review* 49 (Winter 1973): 303–331; Jay B. Nash, ed., *The New Day for the Indians: A Survey of the Working of the Indian Reorganization Act of 1934* (New York: Academy Press, 1938); and Graham D. Taylor, *The New Deal and American Indian Tribalism: The Administration of the Indian Reorganization Act, 1934–1945* (Lincoln: University of Nebraska Press, 1980).

The Indian Claims Commission has generated scholarly investigations, including Thomas LeDuc, "The Work of the Indian Claims Commission Under the Act of 1946," *Pacific Historical Review* 26 (February 1957): 1–16; Nancy Oesterich Lurie, "The Indian Claims Commission

Act,'' *Annals of the American Academy of Political and Social Science* 311 (May 1957): 56–70, and ''The Indian Claims Commission,'' *Annals of the American Academy of Political and Social Science* 436 (March 1978): 97–110; Herbert T. Hoover, ''Yankton Sioux Tribal Claims Against the United States, 1917–1975,'' *Western Historical Quarterly* 7 (April 1976): 125–142; John R. White, ''Barmecide Revisited: The Gratuitous Offset in Indian Claims Cases,'' *Ethnohistory* 25 (Spring 1978): 179–192; and Robert C. Carriker, ''The Kalispel Tribe and the Indian Claims Commission Experience,'' *Western Historical Quarterly* 9 (January 1978): 19–31.

Termination has received considerable recent treatment. Excellent monographs include Larry W. Burt, *Tribalism in Crisis: Federal Indian Policy, 1953–1961* (Albuquerque: University of New Mexico Press, 1982), and Donald L. Fixico, *Termination and Relocation: Federal Indian Policy, 1945–1960* (Albuquerque: University of New Mexico Press, 1986). A comprehensive legal summary is Charles F. Wilkinson and Eric R. Biggs, ''Evolution of the Termination Policy,'' *American Indian Law Review* 5, no. 1 (1977): 139–184. Specific tribal studies include Stephen Herzberg, ''The Menominee Indians: Termination to Restoration,'' *American Indian Law Review* 6, no. 1 (1978): 143–186; Nancy Oestreich Lurie, ''Menominee Termination: From Reservation to Colony,'' *Human Organization* 31 (Fall 1972): 257–270; and Susan Hood, ''Termination of the Klamath Tribe of Oregon,'' *Ethnohistory* 19 (Fall 1972): 372–392.

Relocation policy has been fully explored only in Fixico, *Termination and Relocation*. One of the best law review articles on American Indian law and history is devoted to explaining the complexities of Public Law 280: Carole E. Goldberg, ''Public Law 280: The Limits of State Jurisdiction over Reservation Indians,'' *UCLA Law Review* 22 (February 1975): 535–594.

Efforts to restore terminated tribes to full federal recognition are discussed in William W. Quinn, Jr., ''Federal Acknowledgment of American Indian Tribes: The Historical Development of a Legal Concept,'' *American Journal of Legal History* 34 (October 1990): 331–364; ''The Unilateral Termination of Tribal Status: *Mashpee Tribe v. New Seabury Corp.*,'' *Maine Law Review* 31, no. 1 (1979): 153–170; Michael C. Walch, ''Terminating the Indian Termination Policy,'' *Stanford Law Review* 35 (July 1983): 1181–1215; and Beth Ritter Knoche, ''Termination, Self-Determination and Restoration: The Northern Ponca Case'' (M.A. thesis, University of Nebraska–Lincoln, 1990).

The Indian Bill of Rights has just begun to receive a comprehensive treatment. The secondary literature is confined to law review assessments. Some of the best are Donald L. Burnett, Jr., "An Historical Analysis of the 1968 'Indian Civil Rights' Act," *Harvard Journal of Legislation* 9 (May 1972): 557–626; Arthur Lazarus, Jr., "Title II of the 1968 Civil Rights Act: An Indian Bill of Rights," *North Dakota Law Review* 45 (Spring 1969): 337–352; "Implications of Civil Remedies Under the Indian Civil Rights Act," *Michigan Law Review* 75 (November 1976): 210–235; G. Kenneth Reiblich, "Indian Rights Under the Civil Rights Act of 1968," *Arizona Law Review* 10 (Winter 1968): 617–648; James R. Kerr, "Constitutional Rights, Tribal Justice, and the American Indian," *Journal of Public Laws* 18 (1969): 311–338; Joseph de Raismes, "The Indian Civil Rights Act of 1968 and the Pursuit of Responsible Tribal Self-Government," *South Dakota Law Review* 20 (Winter 1975): 59–106; Dennis R. Holmes, "Political Rights Under the Indian Civil Rights Act," *South Dakota Law Review* 24 (Spring 1979): 419–446; Gary D. Kennedy, "Tribal Elections: An Appraisal After the Indian Civil Rights Act," *American Indian Law Review* 3, no. 2 (1975): 497–508; Mary L. Muehlen, "An Interpretation of the Due Process Clause of the Indian Bill of Rights," *North Dakota Law Review* 51 (Fall 1974): 191–204; Ralph W. Johnson and E. Susan Crystal, "Indians and Equal Protection," *Washington Law Review* 54 (June 1979): 587–631; and Alvin J. Ziontz, "In Defense of Tribal Sovereignty: An Analysis of Judicial Error in Construction of the Indian Civil Rights Act," *South Dakota Law Review* 20 (Winter 1975): 1–58.

Santa Clara Pueblo v. Martinez has generated considerable scholarly debate. See "Equal Protection Under the Indian Civil Rights Act: *Martinez v. Santa Clara Pueblo,*" *Harvard Law Review* 90 (January 1977): 627–636; Alvin J. Ziontz, "After *Martinez:* Civil Rights Under Tribal Government," *University of California–Davis Law Review* 12 (March 1979): 1–35; and Andra Pearldaughter, "Constitutional Law: Equal Protection: *Martinez v. Santa Clara Pueblo*—Sexual Equality Under the Indian Civil Rights Act," *American Indian Law Review* 6, no. 1 (1979): 187–204.

The restoration of lands to tribes has been chronicled in several essays. On the Alaska Native Claims Settlement Act, see Arthur Lazarus,

Jr., and W. Richard West, Jr., "The Alaska Native Claims Settlement Act: A Flawed Victory," *Law and Contemporary Problems* 11 (Winter 1976): 132–165, and Sidney L. Harring, "The Incorporation of Alaskan Natives Under American Law: United States and Tlingit Sovereignty, 1867–1900," *Arizona Law Review* 31, no. 2 (1989): 279–327. On the Maine Indian Claims Settlement Act, see Paul Brodeur, "Annals of Law," *New Yorker,* October 11, 1982, pp. 76–155; Harry B. Wallace, "Indian Sovereignty and Eastern Indian Land Claims," *New York Law School Law Review* 27, no. 3 (1982): 921–950; and Francis J. O'Toole and Thomas N. Tureen, "State Power and the Passamaquoddy Tribe: 'A Gross National Hypocrisy'?" *Maine Law Review* 23, no. 1 (1971): 1–39.

Self-determination is the subject of Kenneth R. Philp, ed., *Indian Self-Rule: First-Hand Accounts of Indian–White Relations from Roosevelt to Reagan* (Salt Lake City: Howe Brothers, 1986). Additional accounts can be examined in Vine Deloria, Jr., "Self-Determination and the Concept of Sovereignty," in *Economic Development of American Indian Reservations,* ed. Roxanne Dunbar Ortiz (Albuquerque: University of New Mexico Native American Studies, 1979), pp. 22–28; Rebecca L. Roggins, "The Forgotten American: A Foundation for Contemporary American Indian Self-Determination," *Wicazo Sa Review* 6 (Spring 1990): 27–33; and Emma R. Gross, *Contemporary Federal Policy Toward Indians* (Westport, Conn.: Greenwood Press, 1989).

Modern Indian Rights Issues

During the past twenty years, many new issues never before litigated or legislated in Indian law have been the center of considerable attention, including rights to natural resources, land-use powers, taxation, control of criminal and civil jurisdiction over reservations, rights of Indian children, religious freedom, gaming, and repatriation. Excellent sources for commentary include Native American periodicals such as *Indian Country Today* (formerly *Lakota Times* [Rapid City, S.D.]) and the newsletter of the Native American Rights Fund: *The NARF Legal Review* (formerly *Announcements*). No full-length treatments of most of these topics yet exist.

Several recent works address Indian rights to their reservations' natural resources. A good survey is Marjane Ambler, *Breaking the Iron Bonds: Indian Control of Energy Development* (Lawrence: University

Press of Kansas, 1990). The Nuclear Waste Policy Act is considered in Nancy E. Hovis, "Tribal Involvement Under the Nuclear Waste Policy Act of 1982: Education by Participation," *Journal of Environmental Law and Litigation* 3 (1988): 45–65. Zoning is investigated in Osborne M. Reynolds, Jr., *"Agua Caliente* Revisited: Recent Developments as to Zoning of Indian Reservations," *American Indian Law Review* 4, no. 2 (1976): 249–267, and Jessica S. Gerrard, "Undermining Tribal Land Use Regulatory Authority: *Brendale v. Confederated Tribes,"* *University of Puget Sound Law Review* 13 (Winter 1990): 349–375.

Taxation has become a very controversial issue. See Daniel H. Israel, "The Reemergence of Tribal Nationalism and Its Impact on Reservation Resource Development," *University of Colorado Law Review* 47 (Summer 1976): 617–652; Russel Lawrence Barsh, "Issues in Federal, State, and Tribal Taxation of Reservation Wealth: A Survey and Economic Critique," *Washington Law Review* 54 (June 1979): 531–586, and "The Omen: *Three Affiliated Tribes v. Moe* and the Future of Tribal Self-Government," *American Indian Law Review* 5, no. 1 (1977): 1–73; Carole E. Goldberg, "A Dynamic View of Tribal Jurisdiction to Tax Non-Indians," *Law and Contemporary Problems* 40 (Winter 1976): 166–189; and Katherine B. Crawford, "State Authority to Tax Non-Indian Oil & Gas Production on Reservations: *Cotton Petroleum Corp. v. New Mexico,"* *Utah Law Review,* no. 2 (1989): 495–519.

The protection of Indian children from non-Indians is the subject of Linda A. Marousek, "The Indian Child Welfare Act of 1978: Provisions and Policy," *South Dakota Law Review* 25 (Winter 1980): 98–115; David Woodward, "The Rights of Reservation Parents and Children: Cultural Survival or the Final Termination?" *American Indian Law Review* 3, no. 1 (1975): 21–50; Gaylene J. McCartney, "The American Indian Child–Welfare Crisis: Cultural Genocide or First Amendment Preservation," *Columbia Human Rights Law Review* 7 (Fall–Winter 1975–1976): 529–551; and Russel Lawrence Barsh, "The Indian Child Welfare Act of 1978: A Critical Analysis," *Hastings Law Journal* 31 (July 1980): 1287–1336.

The American Indian Religious Freedom Act is explored in Howard Stambor, "Manifest Destiny and American Indian Religious Freedom: *Sequoyah, Badoni,* and the Drowned Gods," *American Indian Law Review* 10, no. 1 (1982): 59–89; Stephen McAndrew, *"Lyng v. Northwest Indian Cemetery Protective Ass'n:* Closing the Door to Indian Religious Sites," *Southwestern University Law Review* 18, no. 4 (1989):

603–629; and Peggy Doty, "Constitutional Law: The Right to Wear a Traditional Indian Hair Style—Recognition of a Heritage," *American Indian Law Review* 4, no. 1 (1976): 105–120.

Finally, gaming is discussed in a useful article by Allen C. Turner, "Evolution, Assimilation, and State Control of Gambling in Indian Country: Is *Cabazon v. California* an Assimilationist Wolf in Preemption Clothing?" *Idaho Law Review* 24, no. 2 (1987–88): 317–338. Repatriation issues are just beginning to be published in scholarly periodicals.

Index of Cases

Index